W9-CUJ-214

The Point of Production

Work Environment in Advanced Industrial Societies

JOHN WOODING
CHARLES LEVENSTEIN

THE GUILFORD PRESS
New York   London

© 1999 The Guilford Press
A Division of Guilford Publications, Inc.
72 Spring Street, New York, NY 10012
http://www.guilford.com

Printed in the United States of America

This book is printed on acid-free paper.

Last digit is print number: 9   8   7   6   5   4   3   2   1

Library of Congress Cataloging-in-Publication Data

Wooding, John.
    The point of production : work environment in advanced
industrial societies / John Wooding and Charles Levenstein.
        p.   cm.—(Democracy and ecology)
    Includes bibliographical references and index.
    ISBN 1-57230-447-2 (paper)
    1. Work environment.   I. Levenstein, Charles.
II. Title.   III. Series.
T59.77.W66   1999
658.5—dc21   99-18040
                                                               CIP

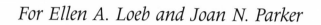

For Ellen A. Loeb and Joan N. Parker

About the Authors

John Wooding, PhD, is an associate professor of political science and Chair of the Department of Regional Economic and Social Development at the University of Massachusetts, Lowell. He has written on occupational health and safety policy in the United States and Great Britain and specializes in international regulatory policy.

Charles Levenstein, PhD, is a professor of work environment policy at the University of Massachusetts, Lowell; an adjunct professor of community health at Tufts University Medical School; a professor of occupational health policy at De Montfort University, Leiscester, UK; and a visiting faculty member in the Department of Environmental Science and Policy at the Central European University, Budapest. Dr. Levenstein has written extensively on occupational safety and health and work environment policy. He is the editor of the journal *New Solutions*.

Introduction to the Democracy and Ecology Series

This book series titled "Democracy and Ecology" is a contribution to the debates on the future of the global environment and "free market economy" and the prospects of radical green and democratic movements in the world today. While some call the post-Cold War period the "end of history," others sense that we may be living at its beginning. These scholars and activists believe that the seemingly all-powerful and reified world of global capital is creating more economic, social, political, and ecological problems than the world's ruling and political classes are able to resolve. There is a feeling that we are living through a general crisis, a turning point or divide that will create great dangers, and also opportunities for a nonexploitative, socially just, democratic ecological society. Many think that our species is learning how to regulate the relationship that we have with ourselves and the rest of nature in ways that defend ecological values and sensibilities, as well as right the exploitation and injustice that disfigure the present world order. All are asking hard questions about what went wrong with the worlds that global capitalism and state socialism made, and about the kind of life that might be rebuilt from the wreckage of ecologically and socially bankrupt ways of working and living. The "Democracy and Ecology" series rehearses these and related questions, poses new ones, and tries to respond to them, if only tentatively and provisionally, because the stakes are so high, and since "time-honored slogans and time-worn formulae" have become part of the problem.   –JAMES O'CONNOR, *Series Editor*

Chapter Two is derived from C. Levenstein and D. Tuminaro, "The Political Economy of Occupational Disease," *New Solutions*, vol. 2(1), 1992, pp. 25–34. © 1992 by the Oil, Chemical, and Atomic Workers International Union. Adapted by permission of Baywood Publishing.

Chapter Four is derived from C. Levenstein, J. Wooding, and B. Rosenberg, "The Social Context of Occupational Health," in B. Levy and D. Wegman (Eds.), *Occupational Health: Recognizing and Preventing Work-Related Disease* (Boston: Little, Brown, 1995). © 1995 by Little, Brown & Co. Adapted by permission of Lippincott Williams & Wilkins.

The section of Chapter Six on fraud relies heavily on the unpublished research of Mary L. Dunn. The section "A Historical Bloc: The Malingerer, the Shyster, and the Quack" was originally included in K. Rest, C. Levenstein, and J. Ellenberger, "A Call for Worker-Centered Research in Workers' Compensation," *New Solutions*, vol. 5(3), 1995, pp. 71–79. © 1995 by the Oil, Chemical, and Atomic Workers International Union. Adapted by permission of Baywood Publishing.

Some of the material on scientific research in the private sector in Chapter Seven was developed with Professors Margaret Quinn and Kathleen Rest for a conference on good practices for occupational research in the private sector.

Permission to reprint the following material is gratefully acknowledged:

Extracts on pages 71–72: © 1997 by Farmworker Justice Fund. Reprinted by permission.

Extract on pages 95–96: © 1997 by Associated Press. Reprinted by permission.

Extract on page 102: © 1998 by A. M. Best Co. Reprinted by permission.

Extract on pages 119–123: © 1997 by *Providence Journal*. Reprinted by permission.

Contents

CHAPTER ONE

Introduction

The most telling and significant changes in the national and international economic order in the past twenty years—the growth of new markets and the disappearance of old ones, new technologies, new competitors, demographic shifts, and shifts in investment—all directly affect production and work. More importantly, they have conditioned and altered political structures, modes of political interaction, and the politics of change in the United States and elsewhere.

During this momentous period the American and European political landscapes have witnessed profound rightward shifts: the rise of conservative forces in the mainstream political system; the ideological movement against state interference and regulation; and direct attacks on workers, unions, and the poor. The degradation of the environment, the cutting of wages, the decline in real incomes for the majority of Americans, and the speedup of work all appear to indicate that market capitalism has triumphed—but not without serious negative consequences.

Internationally, the end of the Cold War and the success of market over planned economies; the development of free enterprise systems in eastern Europe and Southeast Asia; the passage of the North American Free Trade Agreement (NAFTA); and the establishment of the World Trade Organization (WTO) as an organ for the promotion of free trade all seem to indicate that the world is cast firmly in capitalism's mold. In short, the world today seems to reflect the success of neoliberalism.

Nevertheless, an essential truth remains: the system still produces goods and services—in fact, it is doing so at an unprecedented rate. Production still takes place. It may not be in the mine or the blast furnace, but it remains in the factory, the office, and the warehouse, and in vast transportation systems across the globe. Here workers work as they always have, though the products may be very different, the conditions of work infinitely more variable, and the workers themselves more diverse than at any time in history. But work they do. Harder than at any time, for less real income.

That this is the case concerns us. We believe that the rapid pace of change conceals the fact that production still matters. What goes on at the point of production still conditions much else in the world. In essence, then, in this book we argue that the fundamental tenets of a materialist analysis of the forces of production still provide fertile ground for understanding the world, in much the same way that they did for Karl Marx in his analysis of nineteenth-century Britain.

This book is about the *point* of production: what the point of production is for, what it defines, what it creates in the economic, political, and social realms. We are reexamining the basic premises of Marxism through the prism of twentieth-century struggles and how we imagine twenty-first-century workplaces will be. By analyzing the political economy of the work environment, we focus on how our material world is created, re-created, realized, distributed, rationalized, and preserved. We believe that production matters, and more so than at any time in history. And production—what goes on in the factory, the office, and the warehouse—determines much of what goes on in people's heads, how they think about their world, and how they view social relations, hierarchy, politics, and themselves. In this sense we remain convinced, as old-school Marxists used to claim, that the infrastructure still determines the superstructure.

THE POINT OF PRODUCTION

What is the point of production? In traditional terms, "at the point of production" means the place where workers, using capital, fashion raw materials into products: the shop floor. This is where the boss really makes his money, where "exploitation" takes place, where creative, live labor is coined into gold, as they say in the old songs. It is also the place where toxic materials are created or used; where waste—hazardous and the plain old kind—falls to the floor or diffuses into the air and may find its way into the environment; where nature is improved upon, from one point of view, or where it is degraded, from another.

But what is the point of production? As comedian George Carlin has put it, we need "stuff." We need shelter, we need food, we need transportation, we need medicine, we need movies, we need electric guitars—the "stuff" necessary to live decently. We may not need all that is produced; and not everybody gets all they need. Some people hardly get anything. The point of production is not to destroy the earth or our surroundings, nor to dump filth into our rivers or spew toxic substances into the air. The point should be socially necessary production, and there is no inherent reason why humans or environment must be destroyed in the process.

But, also, the point of production depends on one's point of view. The people who own and control the production process do so not in order to make "stuff" but in order to make money. They also do not particularly want to destroy humans or the environment, but it may *absolutely* necessary to do so in order to make money. Or, perhaps, it may be *relatively* necessary to do so because other more expensive ways of producing result in *less* money.

Workers produce goods and services because they have to make a living. Consumers buy stuff that is necessary for their lives. Workers and consumers are mostly the same people, but often they forget this. Workers, consumers, and community residents are also the same people. Workers, consumers, and community residents may also be thought of as men, women and children, as gays and straight, as whites, blacks, Latinos, Russian émigrés, Haitians, Salvadorans, Liberians, Bosnians, and so on. Other interesting classes and fragments of classes may be considered as we travel around the world. Most of these people work to live (rather than vice versa).

This book is therefore ultimately about socialist politics for the twenty-first century. We see in the political economy of the work environment a way of looking at and understanding the world. We believe that by understanding the issues and struggles that occur at the point of production we can identify and analyze both the economic and political determinants of the fundamental structural shifts now besetting political relations in the United States and elsewhere.

Over the past twenty years the fate of U.S. capitalism has hinged on whether it can exploit and expropriate workers and the environment on a global scale. At the same time, it has largely completed its domestic political project of pushing down wages, reducing welfare, decreasing government regulation, and breaking unions and environmental movements. Such a context makes any progressive struggles against the economic and ideological hegemony of capitalism problematic. All of these features are well known. They provide the political–economic setting in which struggles over workplace and environmental contamination take place.

These interactional and domestic capitalist forces are not, however, simply structural. Humans act in the domestic economy, in the market, in the global economy. The capital of multinational corporations may be escaping national control, but there is no magic in the international market; the markets and the corporations are still human creations and are subject to human—political—control. Where resistance is problematic, it is also real. Workers, unions, and citizens continue to struggle against bad working conditions, exploitation, and environmental degradation. Some of these struggles are even successful. We believe that the failures and the successes of worker and community movements against the degradation of work and the environment need to be examined. We also believe that this cannot be done without a theoretical and political framework for comprehending the point of production. Creating this framework is one of the major tasks we are undertaking in this book.

THE POLITICAL ECONOMY
OF THE WORK ENVIRONMENT

The production of goods and services, whether that takes place under a capitalist, a mixed, or a centrally planned economy, engenders costs to workers and the environment. Classical economics views "externalities"—such problems as occupational injuries and disease and environmental degradation—as the consequence of industrial development. These externalities—unless they are readily quantifiable in advance—are rarely, if ever, included in the overall costs of production. Frequently, however, they pose major threats to the lives of workers and members of local, regional, and even international communities. Clearly, technological development, the organization of work, decisions about production methods, and what is produced all have a impacts both within outsider the workplace.

Most existing research on these issues deals explicitly with either the problems facing workers from exposure to toxic substances and dangerous working conditions *or* the impact that such production has on the community. In contrast, we place production itself at the center of these issues in this study. In so doing, we link industrial injury and disease to the wider problem of general public health.

We set out a theory of the *work environment*. Our theoretical insights provide a framework for assessing the roles of key actors, beliefs about the social relations of production, dominant ideological assumptions underlying the role of science in work environment policy, and the impact of global economic conditions on occupational and environmental health. We conclude with an argument for the democratization of work and working conditions and for engaged citizens input into production decisions. Underlying the argument of this book and laid out in Chapter Two is our assertion that the conditions of what we call the "work environment" are typically undertheorized in the literature. No exhaustive effort has yet been made to provide a theoretical framework for understanding the totality of working conditions. In this chapter, drawing on insights from Marxist, ecological, and feminist perspectives, we suggest that the central questions facing those concerned about the consequences to public health of modern technology originate squarely within the point of production. We establish the validity of this theoretical approach and discuss in detail the relationship between technology and workers, management, and health professionals, as well as the social, economic, and ideological context of the work environment.

As we have noted, extraordinary changes have taken place in both national economies and the world economy as a whole during the past two decades, among them: increasing use of technology and automation in the workplace; shifts toward service production in the mature capitalist societies; dramatic changes in international competition; an extraordinary increase in the rate of capital mobility; and the internationalization of consumer and capital markets. In Chapter Three we examine what is meant by "technology" and technological development and how technological changes have resulted in developed industrial economies' facing deep challenges to their economic dominance of world commerce. These challenges have resulted in profound economic crisis. Responses to these problems have varied widely among the various economies, but increasingly the challenge posed by technology (for that is what we believe it is) has resulted—particularly in the United States—in an attack on worker and community rights, stagnant or falling wages and standards of living for the majority, a weakening of unions' power and influence, and an attack on the validity and viability of government's involvement with the market and civil society. These phenomena have also prompted a movement toward "deregulation," which also has the effect of curbing the power of trade unions and, in foreign economies, undermining the rationale for government ownership of manufacturing industries and services.

All these developments have had major consequences in the United States and other advanced industrial societies: the erosion of the middle class, the increasingly unequal distribution of income, the rising incidence of poverty (particularly among children), the transformation of full-time jobs into increasingly temporary and part-time work, the declining availability of health benefits, and weakened social, environmental, and occupational safety and health regulation. Chapter Three examines these phenomena and argues that they are important components in understanding the dynamics of the work environment.

It is clear that, given these phenomena, the pattern of occupational disease and injury in a particular society is affected by the level of economic and technological development, by the societal distribution of power, by gender, race, culture, ethnicity, and by the dominant ideology of a particular social and political system. All of these factors bear on the way in which disease and injury are "produced," whether or not the problem is recognized by relevant actors, whether a problem once recognized is controlled or even eliminated, and the extent to which workers receive medical care or compensation for their injuries.

Therefore, to fully understand the problem of occupationally caused injury and disease we need to examine how these issues impinge on the structure of production, the social organization of work, and on society's response to the issue. Chapter Four examines how rapid technological change has affected the work environment in general and the lives of workers, minorities, and women in particular.

The conditions of the work environment are determined by, among other things, the level of technological development, the social organization of work, the structure of economic development, and the balance of power between workers and managers both within and outside the factory. One key issue, however, is in what ways the conditions of work are determined by regulatory policy. The politics of work environment regulation are complex and imprecise.

The post-World War II increase in the level of governmental regulation raises important theoretical and political questions. The state's role brings the institutions for the administration and direction of society into direct confrontation with the social and economic needs of society as a whole. The resolution of these conflicts and the way in which the state, as a set of institutions, has dealt with these problems are both parts of this story. But, more importantly, it is the relationship between the institutions of state control and the social and economic forces that gravitate around safety and health at work that provides us entrée to a better understanding of the political economy of the work environment. In Chapter Five we provide an overview of the development of regulatory policy in the United States and begin to explore how various actors relate to the work environment, paying particular attention to the role of the state.

Efforts by workers to gain protection from the economic effects of occupational disease and injury affect the definition of disease and injury and subsequent research on occupational health hazards as well as research on and the implementation of control technologies. Scientists may require evidence of an invariable link between an exposure and a claimed health effect in order to put their stamp of approval on the occupational etiology of a worker's illness. Nowhere is this process more critical than in the way the system provides compensation to injured workers.

The issue of workers' compensation is discussed in Chapter Six. By and large, the system is geared toward handling injuries rather than occupational disease. Workers made sick by the work they do must demonstrate the occupational etiology of their illness. The burden of proof is the employee's in instances where employers or insurance carriers challenge whether the illness is really work-related, that is, that its cause originates in the workplace. In this chapter we explore the impact of the importance of economic considerations in the definition of diseases, how the relative political strength of labor and management will determine what is compensable and who gets compensated, and the role that workers' compensation plays in defining occupational ill health.

Injury and disease resulting from production are, of course, crucially linked to the role of science and scientific knowledge. Scientific professionals are deeply engaged in the investigation, understanding, and prevention of occupationally and environmentally produced health problems. But how and in whose interests are most scientific studies undertaken? How is epidemiological evidence about disease causation evaluated? Who funds scientific research? How "objective" is scientific knowledge in the province of work environments?

In exploring these questions in Chapter Seven, we assess the ways in which occupational health research is undertaken. We analyze data on funding for such research, the arguments proposed for preventive measures, and the use of risk assessment in the regulatory process. Our central argument here is that this is a highly political process, reflecting the unequal distribution of power and resources in society as a whole and around the question of research on occupational health in particular.

In this chapter we also discuss the role of professionals within the occupational health setting, presenting an analysis based on the contradictory nature of this role. Scientific knowledge and the application of science to the problems of the work environment do not occur in a vacuum; rather, they are produced within a system of dynamic social relations that must be fully understood to appreciate the way in which science and scientists approach these issues. The question of ethics under such circumstances is, therefore, problematic.

CONCLUSION

Occupational injury and disease are economic phenomena resulting from social decisions about technology and the use of labor in the production of goods and services. The workplace setting presents unique problems for public health because, on the one hand, virtually all the hazards are environmental and can be prevented or controlled while, on the other hand, the work environment is a setting for social conflict with large economic stakes. The rights of property owners (even in state socialist systems), the economic obligations of managers to owners of enterprises, and the imbalance of power between labor and management present particular problems for occupational health. The position of health and safety professionals in industry is frequently problematic because of tensions between their responsibilities to employers and the ethical codes of their professions. The imperatives of production (and profit) frequently override other responsibilities relating to the health and welfare of employees.

The thematic coherency of this work is reflected in the careful articulation of a theory of the political economy of the work environment. A conscious effort has been made to locate the issues surrounding occupational and environmental health within the context of the overall power relations within society. We argue that such a framework provides greater insight because it comprehends these problems as originating within the economic and political structure of the domestic political economy and because attention is paid to the interaction of that economy with global economic forces. We also argue that the ideological context defines the definition, detection, and possible amelioration of occupationally produced disease and injury.

Given the assumptions of this argument, we conclude in Chapter Eight with a discussion of the potential for full democratic control over technological decision making, investment and production decisions, and participation in the generation and evaluation of scientific knowledge.

CHAPTER TWO

The Political Economy of the Work Environment

Work and production may justifiably be regarded as *the* central human activity. The production of goods and services, the satisfaction of needs and wants exist in the material realm of production, whether undertaken in a free-market, mixed, centrally planned, or predominantly agricultural economy. Increasingly the production and distribution of goods is a global phenomenon. Work occupies much of most people's lives, and all are affected by the production process either directly through their participation in that process or indirectly through its consequences or results.

We can assume, therefore, that the conditions of work determine a variety of other political, social, and economic phenomena. Most importantly, of course, working conditions determine the quality of life and the health of those who work at the *point of production,* whether that is the factory floor, the field, the office, or the warehouse. Given this assumption, we define the work environment in the broadest possible sense: it includes the conditions of work primarily but also the consequences of work for the economic and social conditions of the community and the surrounding environment. An understanding, therefore, of what the work environment is, how it functions, and what its consequences are, is central to our knowledge of general working and living conditions.

In what follows we begin to develop an argument that addresses the political economy of this work environment. Our analysis is grounded in the material conditions of production but also embraces the political, sociological and cultural conditions in which production takes place. In analyzing this *work environment* we must begin by establishing what forces and actors are most important in determining the shape of working conditions.

Since we are concerned with the actions and structures created at the point of production, we ground our analysis in the central relationship between the worker and work. The life of workers is absolutely central to the argument that follows. The most important question here is how the work environment creates the conditions that result in occupational and environmental disease and injury. In what ways can we understand the work environment as the primary context for the generation of ill health that affects workers, communities, and the general environment?

A THEORY OF THE WORK ENVIRONMENT

Workplace injury and illness is an unacknowledged epidemic in the United States: approximately 16,000 workers are injured on the job every day, of whom 17 die (crushed by machines, falling from scaffolding, run over by trucks or forklifts, electrocuted, or shot). Another 135 die every day from diseases caused by exposure to toxins and chemicals in the workplace. This toll is the equivalent of a major airplane crash every day in which all passengers and crew are killed (Silverstein, 1995). The economic costs are equally staggering: $173.9 billion in direct and indirect costs, or 3 percent of the gross domestic product (Leigh, Markowitz, Fahs, Shina, & Landrigan, 1996).

As new forms of work multiply, new types of dangers become evident. Ergonomic problems are increasing, and repetitive motion disorders, such as carpal tunnel syndrome, are becoming ubiquitous: 27,000 repetitive motion injuries were recorded by the Bureau of Labor Statistics in 1983, 224,000 annually by 1991. This problem continues to worsen. In 1996 carpal tunnel syndrome accounted for the highest median number of days away from work for major disabling conditions—twenty-five days (U.S. Department of Labor, Occupational Safety and Health Administration, 1996).

In 1997, the budget for the Occupational Safety and Health

11

Administration (OSHA) was $325 million, that for the Food Safety and Inspection Service (the federal agency that oversees the nation's food quality) was nearly $650 million, while the Environmental Protection Agency (EPA) had a 1997 budget of $7 billion.[1] To police the workplace, OSHA had fewer than 1,000 active inspectors. With 6.5 million workplaces in the United States this means that at its current pace OSHA can inspect each workplace once every eighty-seven (Reich, 1993). The federal government, through OSHA, will spend just $3 per worker annually to deal with this problem (Silverstein, 1995).

Nevertheless, it is not just the threats to health nor even the economic costs that such conditions pose (serious as they are), but rather also the fact that many workers typically labor under degrading and even intolerable, conditions. Few fully comprehend the deplorable conditions of work as experienced by many workers in the United States and by many more in the rest of the world—in mines, meat-packing plants, small factories, warehouses, in print shops, auto repair shops, machine tool plants, and maintenance rooms, too often places that are full of dirt and grime, where toilets are filthy, fresh air virtually nonexistent, and eating facilities either absent or unfit for use (Parker & Solomon, 1995).[2] Imagine how a business executive might react to conditions such as these, or what outrage there would be if this were the work environment in corporate corridors and government offices.

It is difficult to be exact about the toll in human suffering of workplace injury and disease in the United States; but the figures that are available are sobering. The United States also compares unfavorably with other industrialized countries, having an incidence of workplace fatalities much higher than in most western European countries (as well as Japan [see Table 2.1]).

Our assertion that this toll of injury and disease is a social construction, not an inevitable consequence of industrial production, is not a new one (Rosen, 1943). Nevertheless, occupational medicine,

---

[1]In addition, the National Institute for Occupational Safety and Health received $141 million, the Mine Safety and Health Administration, $197 million in fiscal year 1997.

[2]One recent example: a small lead smelter, itself a story of the link between occupational and environmental health, provided no lunchroom for its workers. Instead a microwave was perched on top of the toilet tank in the one lavatory. The smelter was completely contaminated by lead dust on all work surfaces. All the workers were immigrants. There was no union.

**TABLE 2.1. Occupational Fatality Rates
in Selected Countries**

Country	Fatality rate*
United States	.105
France	.074
West Germany	.080
Greece	.053
United Kingdom	.034
Sweden	.018
Japan	.030

*Per million person-hours worked. National Safe Workplace Institute, *Unmet Needs: Making American Workplaces Safe for the 1990s* (Chicago: Author); cited in McGarity and Shapiro, *Workers at Risk* (1993, p. 5).

industrial hygiene, occupational epidemiology—and economics—are taught and practiced under the fiction that politics, policy, and social constructs are considerations separate from, and peripheral to, hard science. Much analysis of the work environment can be divided into those works that stress the social and economic factors that shape work, and those that focus on the technical problems confronting efforts to ameliorate workplace hazards (Ashford, 1980; Berman, 1978; Gersuny, 1981; C. Noble, 1986; Donnelly, 1982; Bureau of National Affairs, 1971; Mintz, 1984; Elling, 1986; Robinson, 1991; Dembe, 1996; Garrity, 1995). In particular, there is no systematic exposition of the social, political, and economic factors that significantly shape our understanding of work-related health and safety hazards and their elimination or control.

Occupational diseases and injuries are distinct from other health issues in that they are the *direct result* of economic activity. The analysis presented here places management's control of the workplace, technological decisions, and the labor process at the center of occupational health. This model suggests that the key relationship for understanding the work environment is a "triangle" of control representing management's dominance of the workplace, workers, and any potential hazards. These relationships are embedded in a historical and ideological context, influenced by certain other institutions and individuals, as elaborated in Table 2.2.

Other forces, at certain times and over certain issues, may also play pivotal roles. Activists in the occupational safety and health field can press issues of worker health by providing knowledge and information and by organizing protests. Reporters and journalists

13

may pick up on a particular problem or be prompted by a graphic disaster to investigate the work environment and may thereby also play a significant role in creating awareness about work environment issues. At any one time there may be many reasons why a particular worker health and safety problem gets attention, but the primary actors in the system remain the five key players listed in Table 2.2.

Our discussion of the political–economic aspects of occupational disease and injury will focus on: (1) the production of disease and injury; (2) the perception or recognition of disease and injury; (3) control measures; and (4) compensation for affected workers. To a certain extent these categories are arbitrary since the overriding principles and structure of the political economy provide the context for our discussion. The central principle of profit-making, the debate about the appropriate role of government in industry and the rise of the interventionist state, and the playing out of class and interest group conflict through government are recurrent themes. In the

TABLE 2.2. Key Actors Impinging on the Core Relationship of Worker, Management, and Hazard

1. *Professional consultants, universities, and research institutes* typically provide scientific information about workplace hazards and the means to ameliorate them. The various research institutes may or may not work in collaboration with the government.

2. *The government* typically sets and enforces occupational safety and health standards. In most countries it also plays a key role in providing and initiating research about work environment hazards.

3. *Insurance companies* are also key actors in the United States, as elsewhere. They provide the economic context in which firms obtain workers' compensation and may—by the cost of their premiums—encourage firms to improve health and safety conditions.

4. *Unions* provide collective strength to organized workers. They negotiate working conditions and may provide a counterweight to management's prerogatives. Many unions, in the United States have their own health and safety staff, providing information to workers and workers' representatives. They also put pressure on the government to act responsibly on workplace hazards by lobbying for standard setting, regulation, and enforcement.

5. *The health care system* deals with injured workers. Workers enter that system at various points and with varying degrees of involvement (depending on access).

United States the relative weakness of labor in political and economic terms appears to have influenced the resolution of occupational health and safety issues in certain critical ways, much to the detriment of workers.

THE PRODUCTION OF DISEASE AND INJURY

Occupational disease and injury emanate from the production of goods and services, which involves the exposure of workers to materials, machines, technologies, and work practices that are hazardous to their health and well-being. Our examination involves, in particular, the intersecting roles played by technology, the speed and capacity of production, the structure of an industry, and the nature of the labor market.

Technology and Disease

The choice of technology in production is an engineering and political–economic decision (D. Noble, 1979). The use of particular materials, the organization of work, the employment of particular machines, and their arrangement in relation to one another are subject to economic and social imperatives and constraints. For instance, the introduction of the automatic loom in cotton textile production was, at least in part, the result of labor–management conflict in the industry as well as interfirm competition (Levenstein et al., 1987). The decision to manufacture fluorescent lights, and the resulting hazard of worker exposure to beryllium, was part of a strategy for Sylvania to win an increased share of the market (Zwerling, 1987). Implicit in the search for efficient technologies of production are the requirements to minimize costs and enable management to effectively control labor processes. Employer decisions on the choice of technology seem rarely to consider the full costs of subsequent occupational health and safety hazard(s). Current efforts to remedy dangerous work situations may involve expensive retrofitting of equipment, reorganization of production and difficult retraining of labor, or the unknowable costs of technology-forcing innovation. The sudden profusion of carpal tunnel syndrome problems in recent times as a result of widespread keyboard use represents a major negative health consequence of technological innovation.

The Level of Production

At times, the scale of operations or the intensity of use of equipment or labor can have a substantial impact on hazardous exposures. Increased production requirements for World War II, for instance, may have produced greater dust exposures for asbestos or cotton textile workers (Levenstein et al., 1987). Increasing shift-work may lead to accelerated depreciation of equipment and the failure of dust controls. Particularly susceptible people, ordinarily barred from employment, may be utilized during periods of high demand. Child labor legislation, for instance, is relaxed during wars (Felt, 1965). Moreover, industrialization may increase the sheer numbers of workers exposed to dangerous environments (Legge, 1920).

Unemployment and underemployment, of course, have their own negative impact on health; the risk of hazardous employment may appear attractive to workers faced with debilitating poverty. The net result of industrialization may be a change in the pattern of the cases of morbidity and mortality among workers (Kuhn & Wooding, 1993, 1994).

The Impact of Industry Structure

A more subtle aspect of the generation of occupational disease and injury involves interfirm competition and political–economic definition of industries. Vertical integration in an industry tends to be limited by the potential for stabilizing and planning production; service stations and dealers, for instance, are highly competitive small operations and are not considered part of the highly concentrated automobile industry. Similarly, agriculture is a relatively competitive sector compared to the more highly concentrated industrial users of "raw" materials. Decisions concerning production methods and technology made in one sector of the economy may have grave implications for occupational hazards experienced in another. For example, the mechanization of cotton picking resulted in a higher trash content in the dust of textile mills, probably increasing the incidence of byssinosis (chronic bronchitis, sometimes complicated by emphysema or asthma) experienced by mill workers. On the other hand, the industrialization of agriculture and the development of cash crops for emerging industries in developing countries may change the nature of hazards experienced by farm workers. The generation of such hazards may go unrecognized even in countries where industrial development is planned because of the narrow focus of development strategists (Levy & Levenstein, 1990).

The Nature of the Labor Market

The relative scarcity of labor as a whole or of particular types of labor will influence not only technical choices in production and the resultant hazards associated with laboring in particular industries, but also the options available to labor and the pressures that it can exert on management to improve working conditions. Low levels of employment may limit the influence of trade unions. In some industries, employers, eager to maintain an adequate labor supply and prone to pay more attention to occupational hazards during periods of high employment, may relax standards during periods of declining demand. Further, in the face of particular labor shortages, employers may introduce technical changes entailing unknown hazards in order to reduce costs and undetermine the influence of skilled workers. Finally, attempts to control the labor process may prompt employers to institute technical and administrative changes in production methods that have adverse consequences for the health of employees.

THE PERCEPTION OF OCCUPATIONAL DISEASE

On the one hand, occupational disease and injury are generated by technologies employing labor in production. On the other hand, while ill health may appear to be an "objective" matter, political–economic considerations are important in the perception of disease and injury. Observers of worker health occupy different, and sometimes opposed, locations in the system of production. Therefore, they bring to their understanding of occupational disease and injury viewpoints directed or constrained by other, sometimes more important, determinants of their positions (Bayer, 1988), for instance, managers concerned about absenteeism may focus on employees' lifestyles rather than workplace hazards. Some investigators of occupational health may have conflicting institutional responsibilities. Other observers may have such a peripheral connection to the system of production and its control that they may be blind to the industrial etiology of disease and injury. For example, it may not occur to a dentist to ask about working conditions of patients.

Workers

Presumably, workers themselves are most likely to be aware of their own occupational hazards and their ill health where the relationship between exposure and disease and injury is clear-cut. Occupational health literature is replete with examples of workers alerting health professionals to hazardous situations (Levy & Wegman, 1988). However, the increasing complexity of industrial production and the insistence of employers on control over the production process prevent workers from having information about more subtle, though sometimes devastating, hazards. Workers rely more and more on managers and health professionals to interpret disease for them. Thus, although a worker may "know" he or she is sick, the occupational etiology of the illness may be uncertain or only suspected.

The economic condition of workers may act against their "knowing" the source of their illness. Maintaining family income may be of much higher priority to a worker than the correct diagnosis of the source of an illness. Indeed, if a worker suspects a health and safety hazard on the job and yet perceives no reasonable employment alternative nor the possibility of changing the work environment, the worker may not want to "know" about the hazard. Further, workers may believe that some hazards and illness are simply part of the job—"always have been and always will be"—and therefore do not "know" they exist or acknowledge that such conditions are "abnormal." The realities of working-class life place primary emphasis on the economic survival of the family; thus, the hazards afflicting a breadwinner may be accepted fatalistically (Nelkin & Brown, 1982)—or "suffered gladly."

To the extent that working people depend on management and/or health professionals for interpretation of their illnesses, the foregoing considerations contribute to worker acceptance of assurances that the work environment is "safe." In addition, the cost of health care is often substantial enough to discourage workers from using the health care system.

Workers will mobilize around occupational hazards and perceived disease: (1) if the hazard is believed to be drastic and the effects serious; (2) if clear alternatives seem to be available; (3) if workers' bargaining position vis-à-vis employers seems to be strong; or (4) if new avenues for correcting hazardous situations seem to be available. These situations more often apply in exposures to industrial injuries where the danger is obvious (an unguarded press, poor scaffolding, etc.).

Workers in trade unions may or may not find the unions to be responsive to their health concerns. Some unions are highly attentive to industrial hazards, while others seem fearful that worker militancy about such issues might make political insurgency more likely; yet others may promote interest in occupational health as part of an overall bargaining strategy, to be traded off against other demands. To the extent that the union's leadership believes that managerial control of technological choices is inviolable, bargaining strategists may end up directing members' attention to other, more "winnable," issues.

Finally, it should be noted that worker expectations about their own health might be quite low. To the extent that ill health is presumed, worker attention to occupational hazards will be negligible. Indeed, some workers, particularly low-wage, minority workers, may feel that they may be able to get their jobs only because of the dangers involved and may be fearful that any attention to "cleaning up" the workplace will cost them their employment.

Management

Management's perceptions of worker ill health and occupationally derived disease and injury must be strongly conditioned by economic considerations. In certain circumstances, employers may be concerned with the toll of hazardous exposures as expressed in absenteeism and general debilitation of its work force, particularly if production is affected. In addition, management may worry about a public outcry and/or media attention to industrial hazards, in part because of its fears of government regulation or other public intervention and in part because of concern for "image" and public relations effects. In addition, the cost of workers' compensation insurance coverage can be greatly affected by casual managerial attitudes toward worker health. Finally, management attention may be drawn to health issues because of collective bargaining pressures or, even where there is no union, because of difficulties in recruiting employees to hazardous occupations, or finally because of the possibility of employee unionization as a consequence of health-related grievances.

Nevertheless, there is a great incentive for employers to "externalize" the costs of occupational disease and injury by recognizing workers' ill health but then attributing it to personal habits or to the community environment. Employers, for instance, may be supportive of community efforts to deal with lung disease, such as tuberculosis or smoking-related disease, while denying the existence or seriousness of occupational exposure to respiratory hazards.

Competition alone may engender more attentiveness from some firms than others. A technologically "progressive" firm, which may promote government regulation of exposures at the expense of its more backward competitors, gains a better public image and can better justify its own investment policy in public health terms. Thus, employers that seriously attend to worker health hazards have plenty of economic justification, viewed in the longer-term interests of the corporation.

On the other hand, workers' compensation payments may be regarding as a "dead-weight loss" to many employers. A firm that supports control of exposure to prevent disease and injury may, nevertheless, dispute individual workers' claims for compensation, thus accepting the statistical existence of disease but challenging specific cases (Barth & Hunt, 1980).

Other social factors may affect management's perception of disease and injury. Symptoms may be interpreted as being the result of the poor health status of "inferior" racial stock or of lower class social behavior. The "Monday morning feeling" characteristic of byssinosis, for instance, has been attributed by some to workers' drinking habits on weekends (Levenstein, Mass, & Plantemura, 1987). On the other hand, Alice Hamilton, one of the most important early professionals working in the occupational health and safety field, was able to draw management's attention to occupational disease partly because of her upper-middle-class standing and connections (Hamilton, 1943).

Health Professionals

A critical notion in understanding the perceptions of health professionals is that of "occupational role" (Nowotny, 1975). On one hand, physicians, nurses, and industrial hygienists are trained in scientific, medical, and public health disciplines; on the other, they earn their incomes in a variety of locations relative to the system of production. Frequently, the first indications of a hazardous industrial exposure arise in contact between the worker and his/her own physician, or the community-based health care delivery system. Outside of the company-dominated, single-industry town, however, primary health care personnel have no control and little access to the workplace. Perhaps for this reason, physicians and nurses receive so little training in occupational health and generally give short shrift to occupational considerations in their practice (Levy & Wegman, 1988). When primary care personnel do attend to industrial hazards, they may be constrained in their advice to workers by the limited opportunities for employment that many workers face. Further, physicians specializing in workers' compensation cases deal mainly with injuries and, in any event, are not directly concerned with the controlled elimination of exposures but rather with the compensability of the injury. They have at least indirect access to production decisions—presumably their findings would filter through to the management through experience rating of insurance premiums or through direct controls asserted by insurance companies. Nevertheless, such physicians, since they are in the employ of economically interested actors (that is, insurance companies), tend to discount ill-defined or difficult to diagnose industrial disease. Similarly, physicians and nurses employed by companies have conflicting demands placed on them by their professional responsibilities and their economic dependence on employers. Additionally, the long-term financial viability of the firm may require that attention be paid to hazards that, which costing money to remedy, pose a far graver long-term risk for profitability of the firm. Therefore, the variety of pressures affecting management similarly influence the company "doc" and the industrial nurse (Bayer, 1988). The emphasis in such practice is frequently on preemployment physicals, "loss control," reducing workers' compensation expenses, and ensuring a level of health commensurate with economically functioning employees. Recently, "health promotion in the workplace," focusing on personal behavior rather than occupational hazards, has gained the attention of companies and health professionals concerned about rising medical care costs (Kotin & Gaul, 1980).

Industrial hygienists are generally more closely related to production in the firm, although, just as with company physicians and nurses, they are usually located in staff, rather than line, departments in the corporation. This fundamental distinction is significant, since staff departments are essentially advisory to the management of production, although OSHA and other regulatory agencies may have elevated the importance of health and safety-related professionals in the corporate hierarchy. The training of industrial hygienists, as compared to that of medical professionals, is much more closely related to engineering disciplines, including both the written and unwritten economics of cost-efficient production.

Health and safety personnel, reflect at least three different orientations to the perception of industrial disease and injury (in addition to their in-house loyalties based on economic dependency). The engineering ethic heavily emphasizes efficiency in production. The scientific ethic upholds a conservative approach to information about health effects, leaning toward the requirement of conclusive proof of disease related to particular occupational exposures. The clinical medicine or public health ethic focuses on the condition of the worker, emphasizing preventive measures where there is an indication that injury may be occurring. Thus, economic concerns influence the perception of worker illness through the interaction of occupational roles and professional orientation, depending on the particular profession involved and the location of the particular professional in the scheme of production.

The State and the Legal System

Understanding how the perception of occupational illness and injury is mediated by the state and the legal system requires focus on three concepts or characteristics of modern government in capitalist society: hegemony, class conflict, and bureaucracy. The structural idea of "hegemony," or capitalist dominance, suggests that through the state and its concomitant legal order the dominant class successfully presents itself as the guardian and guarantor of the interests and sentiments of the whole society, including subordinate classes (Alford & Friedland, 1985; C. Noble, 1986; Gramsci, 1971). A somewhat crudely instrumental theory would reduce the role of the state and the rule of law to that of merely serving the interests of the dominant capitalist class, of intervening in social–economic matters in order to mediate and contain class antagonisms in a manner that preserves the legitimacy and stability of the system. Social problems, such as occupational disease and injury, that principally affect the working class would accordingly be downplayed and defined as marginal to the fundamentally satisfactory existent class relations. Recent history, for example, reveals a marked tendency by business to contest occupational disease and injury claims filed under state workers' compensation laws. While the citing of scientific uncertainty concerning many issues about occupational disease does provide a putative legal basis for such controversy, the actual impact of this tactic is to shift the economic burden of occupational disease and injury from capitalists, who exercise control over the work environment, to workers, who already bear the physical and social costs of the disease. Nevertheless, if the state and law can be seen as instruments of class power, they also must be understood as central arenas of class conflict.

Even if the state and its legal system function to mediate and reinforce existing class relations while serving ideologically to legitimate and mystify dominant class power, there are limits and constraints. It is inherent in the nature of the hegemonic state and in the forms and traditions of law that they cannot be reserved for the exclusive use of the dominant class. In the history of occupational health, the black lung movement provides an example of the exercise of counterhegemonic power in its successful attempt to impose its own perception of this occupational disease in the broader medical setting (Smith, 1981). The coal workers struggled for a definition of their illness, fundamentally a political conception, which reflected their collective experience as victims of a preventable disease and injury resulting from political–economic decisions taken by the coal industry. The specific political–economic weight of their organized actions, both legal (lobbying) and extralegal (wildcat strikes), forced the enactment and administration of law responsive to their interests.

Significant class conflict and its reverberations in the state and the legal system are currently evident in respect to compensation and prevention of asbestos-related disease. Although claims are contested and compensation denied, claimants and their lawyers elicit from the legal system a recognition of disease and injury and some degree of remedy through resort to the social welfare laws and through third-party tort liability suits against manufacturers. Even as statutes of limitations barred many claims in certain states, lawyers for claimants persuaded courts to adopt "discovery rules" while legislators introduced amendments to afford some remedy to workers (Barth & Hunt, 1980).

Similarly, the perceived failure of the legal regime to prevent occupational disease and injury by the imposition of liability under compensation laws and tort law (so-called market deterrence) led to regulation by the federal government in the form of the Occupational Safety and Health Act (OSHAct), which was passed in 1970. The establishment of the National Institute for Occupational Safety and Health (NIOSH) under the act constitutes a potentially significant institutional structure that may affect the perception of occupational hazards and disease. It can be seen as one more legal outcome of class conflict and of partially successful struggles by subordinated classes to extract concessions and impose inhibitions on the exercise of power by industry.

Of course, perceptions of occupational disease and injury, which have developed as legal outcomes of class conflict, are not immutable by virtue of being embodied in law. During periods when the underlying political–economic climate is concerned with fiscal austerity, cost effectiveness, and deregulation, erosion of such gains can and does occur. Thus, the extent that the state in modern capitalist democracy constitutes a central location for class conflict, victories, losses, and political compromise is reflected in the official and unofficial stance of regulatory agencies and the legal system in general.

The state and its legal system, however, are not directly administered for the most part by capitalists, but by a middle-class stratum of professionals and bureaucrats. This circumstances raises the issue of the impact of bureaucracy and of bureaucrats, with their particular and distinct political, economic, and professional interests. This arrangement may seem to be an aspect of the foregoing theme of class conflict, and to a certain extent it does embody the variety of considerations included in our earlier discussion of "professional" perceptions of disease (pages 20–22). It is here, however, that we may most usefully explore the significance of a "public health" or "professional" ethic as a distinct political force not entirely dependent on orthodox conceptions of class conflict.

Lawyers and other legal professionals, including judges and legally trained government bureaucrats, have played a somewhat autonomous and highly significant role in working the state and the legal system a central arena of conflict between the fundamental classes. Through adjudication they mediate and legitimate the contending perceptions of industrial disease and injury. Such professionals who administer "the law" come to believe in the long tradition in which they have been steeped, in the particular forms and character of law, its own independent history and internal logic of evolution. They have a profound personal and professional stake in maintaining the apparently neutral stance of law as a body of objective rules applied as logical criteria with reference to standards of equity and universality. Thus, the implementation and shaping of occupational health and safety laws and perceptions may take turns that are reflective of neither management nor labor, but may rest rather on the specific political weight or processes of the bureaucracy. The significance of this bureaucratic, legal, quasiscientific coterie of public servants in an occupational health subsystem of capitalism should not be underestimated.

THE CONTROL OF DISEASE AND INJURY

Political–economic influences on measures to control disease and injury are of no small significance since the introduction of controls raises questions about the appropriate allocation of scarce public and private resources. Further, the choice of control measures may reflect underlying notions of private property rights, the proper role of the state, and the appropriate social control of technology. These are highly politicized choices in the current climate of debate over regulation and deregulation of industry.

Industrial Hygiene

Industrial hygiene approaches involve the elimination or reduction of hazardous exposures to workers through three fundamental measures: engineering controls, personal protection, and administrative measures. Although the use of these "tools" appears to be a technical matter, there is no question that economic considerations are important. Engineering controls may involve substantial capital expenditures and may be linked to the overall investment policy of the firm. Personal protection frequently appears to be less expensive in the short run but may involve substantial maintenance costs over the long haul. Administrative measures, such as job rotation and worker education, sometimes appear to be relatively inexpensive ways to deal with seemingly minor hazards. On the other hand, financial costs may not be the only consideration for firms; the degree of control over the choices of technology and of labor processes may be a conditioning influence for both labor and management (Braverman, 1974). Administrative measures involving worker participation—for example, labeling of materials and worker surveillance schemes—have been attacked by management groups on the basis of cost, but the real opposition seems to be founded on concerns over control.

A basic economic issue in the choice of industrial hygiene controls involves the appropriate way to figure the cost of future expenditures. Future costs must be discounted by an interest rate indicating the return on alternative use of funds. Therefore, current expenditures may loom large compared to ones that will be made in the future. This type of calculation, of course, leads management to be wary of engineering controls, given their attendant large capital requirements. The discount rate chosen by the firm may differ quite substantially from a social discount rate. In any event, any choice may involve guesses about quite distant future economic conditions (Boden, 1979).

Clinical Approaches and Epidemiology

Clinical approaches to the control of occupational disease are most strongly influenced by the sharp separation of primary health care and occupational medicine. The lack of access to the workplace results in a focus on removing the worker from the exposure, rather than eliminating or reducing the exposure. When the disease is not specific to an occupational hazard, the clinician frequently will ignore occupational histories and focus on personal habits (smoking, for example) and therapy for the disease instead. Particularly where employment opportunities are limited, workers may receive medical care for occupational disease with the issue of removal from exposure never even arising. Social class, racial, and ethnic considerations may also be relevant to the situation, particularly when physicians also observe undesirable worker behavior, such as excessive consumption of alcohol, smoking, or "malingering" (Berman, 1978).

Epidemiology's roots in medicine raise related problems.

Because investigators from outside the firm and industry conduct most epidemiological studies, the characterization of exposure may be inadequate. Such statistical studies can only be suggestive of hazardous situations. Since control of exposure data still rests in the hands of management (when the data exist at all)—as is the control of the exposure itself—the effectiveness and value of epidemiological studies rest upon the cooperation of management. Economic concerns thus have an indirect way of influencing studies of occupational disease.

The Importance of the Technical and the Economic

Perhaps the key underlying thread of political economy constraining industrial hygiene and clinical–epidemiological approaches to occupational disease and injury is the nature of the hypothesis that each of these disciplines sets out to test. The existence of disease and injury and warranted controls is examined in a context that is taken as given. Yet the particular economic and technological settings are determined by particular political and economic arrangements and are neither inevitable nor sacrosanct (Navarro, 1982). The leeway afforded to industrial hygiene appears to be a bit wider than that to the medical profession because of the former's more intimate connection with the system of production. Still, the outside health professional may be freer to express concern or alarm about occupational disease and injury in that a peripheral position imposes fewer constraints.

The Political Economy of Regulation

Our earlier discussion of the role of the state (page 22) applies even more aptly to control issues than to disease and injury recognition. Limits to regulation are imposed (at least in the United States) by the imperatives of the capitalist firm, though it is not completely clear how far those limits can be pushed. Economists are especially sensitive to the devastating consequences of regulatory efforts for the capitalist firm. All the same, most firms seem to adjust reasonably well to new regulations, and some even manage to increase profits with new, less hazardous plants and equipment (Ives, 1985). Nevertheless, regulators must take into account the impact of new standards on industry, which means that the fundamental injunction to maximize profits will be protected.

At the same time, the rule-making process has become a new arena for conflict and compromise between economic and political actors. Beyond that, the struggle over enforcement of the rules provides a further battleground for labor and management. Well-organized political forces marshal their experts for these battles. Those who bewail the adversarial nature of health and safety regulation are probably not doing as well as they had hoped in the controversy.

The Political Economy of Research

Occupational health research has been affected and shaped by political–economic factors in a variety of ways:

1. Funding sources may influence the choice of problems to be investigated and the direction of research. Research priorities may be strongly influenced by political and economic actors.

2. The definition of the problem to be investigated may reflect economic interests.

3. The scientific disciplines to be involved in the research may reflect extraneous political considerations, depending on the relative political power or current popularity of particular disciplines.

4. The range of problem solutions considered may be subject to political and/or economic constraints.

5. Conceptions of efficiency in research methodology—reflecting economic constraints on researchers—frequently influence the conduct of research.

6. Political and economic considerations may influence the publication of research findings. The occupational location of researchers—whether in private industry, government agencies, research institutes, or universities—can inhibit the flow of scientific information.

Compensation for Occupational Disease and Injury

Attempts by workers to gain protection from the economic effects of occupational disease and injury affect the definition of disease and injury and subsequent research on occupational health hazards as well as research on and the implementation of control technologies. Scientists may require evidence of an indisputable relationship between an exposure and health effect in order to put their stamp of approval on the occupational etiology of a worker's illness. Much more rigorous definitions of disease develop for occupational illnesses because of the issues of financial compensation than for the "ordinary diseases of life," which involve no employer liability (Barth & Hunt, 1980). Medical researchers are unhappy about "subjective" reports of symptoms in the case of occupationally related illness and search for "objective" technologies by which to make distinctions between disease *caused* by a particular exposure as opposed to illness merely *associated* with particular working situations. Smoking, as non-work-related personal behavior, becomes exaggerated because of the concerns of employers about financial liability.

Principles of Workers' Compensation: Risk

At least in part, the function of workers' compensation insurance programs is to expedite financial relief to workers who experience occupational injury. Rather than permitting (or requiring) that injured employees take their employers to court, workers' compensation is a no-fault insurance program, providing some level of income maintenance and medical benefits. On the other hand, workers' compensation enables employers to avoid potentially disastrous damage suits and to insure against economic liability for occupational hazards. By and large, the system is geared toward handling accidental injuries rather than occupational disease, although some states' original legislation provided for benefits to workers with employment-related illness. Ordinary diseases of life, however, are not compensated. Thus, although workers' compensation is supposed to be no-fault insurance, in the case of disease, employees must demonstrate the occupational etiology of the illness. This requirement can be extremely difficult to meet when the illness is not recognizable as specific to a particular exposure or where the illness takes years to develop, as is frequently the case. The burden of proof is the employee's in instances where employers or insurance carriers challenge the occupational origin of the illness. Financial considerations, then, are extremely important in the definition of diseases potentially compensable under this system. The political strength of labor and management is brought to bear in controversial situations, because a disease may be listed as specifically compensable by state legislatures or workers' compensation commissions. Similarly, political weight can be brought to bear in the selection of commission members.

On the other hand, if workers bring no claims for compensation, the occupational etiology of an illness may go unrecognized. Disease may go untreated or treated by medical practitioners as an ordinary disease of life.

The Uses of Medical Uncertainty

There may be medical uncertainty about the etiology of specific occupational diseases for a number of reasons:

1. Safe conditions may be presumed incorrectly, as was true of the pre-1960s cotton textile industry in the United States. The resulting illnesses of workers may go unrecognized for years or even decades.

2. Workers may not file compensation claims, in which case there may be no particular incentive for scientific research.

3. If there is no clear impact on workers' productivity or if an ample supply of replacement labor is available, employers may find it economically safest to discourage) or even squelch scientific research.

4. The research problem may be difficult—that is, research may require resources and attention beyond the financial capabilities of the interested parties.

5. On the other hand, if substantial economic interests are involved, a plethora of research may be generated, resulting in the obscuring of the issues in question.

6. Finally, neither epidemiology nor medicine is an exact science.

Eventually, expert medical witnesses enter into the politically and economically loaded workers' compensation setting and are asked to testify as to the work-relatedness of claimants' illnesses. Different constructions of information about a disease and about the nature of workers' exposure frequently are possible. Political and

economic bias may interact with the interpretation of medical information.

Compensation and Prevention: Incentives

The workers' compensation system is presumed to encourage preventive measures for the elimination or reduction of occupational
hazards through experience ratings of premiums. Yet, the insurance
aspects of workers' compensation enable employers to *plan* their
expenditures, preventing cataclysmic financial burdens, and thereby
discourage safety measures. In addition, the successful political
efforts of employers to minimize workers' compensation payments,
while preventing suits, further reduce safety incentives.

The emphasis management and labor place on compensation
issues also greatly detracts from efforts to develop useful prevention
information. Workers may press for a broad definition of disease in
order to ensure that sick workers get income maintenance—but
such a broad definition may only serve to mask the specific causes of
disease and deflect research efforts from preventive measures that
may require targeting of specific harmful agents. (Economics, again,
may be a consideration in this discussion of prevention, since cost-
effective control techniques may require knowledge of the specific
agent.) On the other hand, employer resistance to the costs of compensation may result in research that obscures specific agents or situations in order to establish the disease as "ordinary." In any event,
the debate about compensation and its potentially large costs often
keeps attention from being directed to preventive measures. In some
instances, the debate shifts scientific attention to the influence of
smoking, again pulling resources away from controlling the occupational environment. Finally, even though a political resolution of a
compensation controversy might be successful, the pressure for further investigation of the disease could easily wane.

CONCLUSIONS

We have presented a series of hypotheses concerning the social
determinants of occupational disease and injury and, by extension,
their impact on the general environment. They are the key questions
one must ask in analyzing the social history of occupational health,

in studies of particular diseases, in cross-disease comparisons, in related institutional studies, and for cross-national comparisons. In the remainder of this volume we develop and elaborate on this central theory of the work environment, addressing the various actors in that environment and their specific contexts. In the chapters that follow we analyze the role that technology plays in shaping the work environment; how social and political factors interact in constructing the problems of occupational safety and health; the role played by the regulatory state in defining and regulating the work environment; and, how the workers' compensation system and occupational health science define and assess occupational disease and injury.

CHAPTER THREE

Technology and the Work Environment

Standing at the center of any discussion of globalization of capital, and corporate (and governmental) strategies in an era of multinationals, is the work environment, the place where labor power and materials drawn from the natural environment are shaped into goods and services that are then supplied to markets in exchange for other items of value. It is the labor process, as invested in and with "technology," that fundamentally shapes the work environment.

> An essentially human phenomenon, technology is thus a social process; it does not simply stimulate social development from outside but, rather constitutes social development in itself: the preparation, mobilization, and habituation of people for new types of productive activity, the reorientation of the pattern of social development, the restructuring of social institutions, and, potentially, the redefinition of social relationships. (D. Noble, 1979, p. xxii)

And the choice of technology, in the hands of the owners and managers of capital, is the same as the choice of worker injury and disease: it is the other side of the coin, it is what "consumers" need not think about. Thus, the investigation of occupational safety and health is the investigation of alternative industrial technologies; the true "risk factors" are the forces that drive the production managers to choose one or another of the alternatives available, or to invest in others.

Textile production provided the foundation of the industrial revolution in England and the United States. In New England, textile mills and the production of cloth for sale in domestic markets marked the first and most important stage of economic growth in the early nineteenth century. The story of textiles and textile production is well known, as is the role of technology in shaping the nature of that industry. Less well known, and poorly understood, is the relationship between the competitive pressure for more productivity, technological development that improves productivity, and the technology-forcing aspects of government regulation to create a better work environment. Exactly what set of circumstances might arise in which the availability of productivity-enhancing technology would also be a "cleaner" technology? What if a foreign manufacturer were to provide such technology? What is the relationship between production for the domestic market, production for export, and the openness of borders to free trade? To answer these questions we return to essentials. Engels put it succinctly over 100 years ago:

> The materialist conception of history starts from the proposition that the production and, next to production, the exchange of things produced, is the basis of all social structure; that in every society that has appeared in history, the manner in which wealth is distributed and the society divided into classes or orders is dependent upon what is produced, how it is produced, and how the products are exchanged. (Engels, 1976)

WHAT IS TECHNOLOGY?

Technological progress is never without profound human consequences. Too often our perception of technological change fails to capture the immediate impact it has on people's lives. Take the case of an administrative assistant working in a library who, in the early 1990s (1993, October 5–6; anonymous personal electronic communication to author Levenstein), suffered from RSI (repetitive strain injuries—musculoskeletal disorders resulting from repetitive motion, mechanical stress, awkward posture, and local vibration). She had surgery for bilateral carpal tunnel syndrome soon after she was diagnosed, and went on workers' compensation. She had surgery a second time within 6 months, and described the change in her lifestyle as a result of her reduced income (on workers' compensation) and the loss of use of her hands while recuperating from surgery as dramatic. Three years later, though she was able to work, she continued to be in great pain. She believed that her ability to tough it out was due to force of will, not to any lack of severity of her illness. She said her husband still had to help her with all of her household tasks.

She described her job, as an administrative assistant in a library, in which she ordered books and other materials. She used computers for 7½ hours a day, uploading data from a database and bringing it into the library's cataloging system. This involved using two computers, often at the same time. She described the RSI as having affected everything from brushing her hair to more complex things. She said she wished she could take a second job, but was unable to do so because of the pain she was in after work. She had had to abandon hobbies that involved doing work with her hands. Of the things she had to give up, what she said she missed most was being able to hold a book in her hands for any length of time. She described that before, she had lain in bed reading for many hours at a time, but that now she had to sit with a book resting on a table. This caused her to give up reading for pleasure.

After her ordeal, she had had no alternative but to go back to the same job, for approximately $13,000 a year. The job was even more difficult for her physically than it had been before the surgery. She said that because it involved her hands and her arms, RSI had negatively affected every aspect of her life.

We start with what is, at first glance, a rather simple question: what is technology? Defining technology and its relationship to social and political forces becomes, perhaps, the most important task in understanding the work environment. To answer this question we need to go back to some basics.

"Technology" conjures up images of sophisticated computer systems, exotic electronic boxes, incredibly complex weapons systems, the space shuttle, and the like. If we think a little more deeply, oil refineries, nuclear power stations, and rooms humming with computerized equipment come to mind. In short, our image of technology has a very physical form. However, this really limits our understanding of what technology is.

> "Technology" does not just refer to the physical form, the pieces of metal, the electronic components, the chemical compound. Technology refers to the way in which the parts are organized, through the application of knowledge, to realize a particular purpose. (Street, 1992, p. 8)

Underlying this *particular purpose* is a contemporary presumption that technology brings with it inevitable progress, a forward motion that somehow embodies the good life for all. Since the Enlightenment this view of technology, science, and rationality as a kind of trinity that automatically creates a viable future has been the dominant ideology. Despite many doubts about this view, especially since World War I, and despite postmodern critiques, most of our mainstream social and political institutions remain deeply infused with this conception. That this is so is important. The dominance of faith in technology not only blinds us to the many problems technology creates, but also obscures the origins and functions of, and—most importantly—how choices are made about, technology. This is, as we have shown, profoundly important to the work environment.

How then does the issue of *choice* become so important? Who decides what technology to develop and employ is critical because technology is *relatively autonomous*; that is, it builds upon itself at the same time that it is directed and initiated by the decisions of human agents. Technology is both independent and controlled by human activity. The question of the relative autonomy of technology bears many similarities to the discussion in much of the Marxist literature on the autonomy of the capitalist state. In what ways does technology embody prior conflicts? How can technology be said to be independent of human activity and human decision making? Are there points at which technology can be said to reflect the dominance of a particular constellation of social forces, or of class control? To understand more fully, we must briefly enter into a discussion of the various views of the relationship between technology, technical change, and political and social forces. These views appear to be diametrically opposed.

One view is that technology is the causal force in change, a primary force that shapes all other social, economic, and political institutions. Technology drives society. Technology, if you will, is the uncaused cause of human development. As such it can only be responded to. Control is often limited, choice constrained.

The second, opposing version of this argument sees technology as the product of social and political activity. Humans make choices about what technologies come into existence, where and when they are applied to production, and how they are likely to transform social relations. The exercise of power (political, economic, and social) structures the way technology develops. In this perspective, the political economy creates specific technological developments and technological processes. There is no "uncaused cause"; rather, human agents make the critical decisions about technology.

These two opposing views, broadly speaking, ask the same question, namely, does technology cause politics or politics technology? The answer is vitally important to our discussion of the work environment, because we identify the problem of occupational health and safety as arising from choices made about what and how to produce goods and services, and the consequences such choices have for the health of workers. However, it is a mistake to portray these perspectives as being diametrically opposed. The counterposing of politics and technology is, we argue, a false dichotomy. What is really at issue is the relative autonomy of technology from the system of political economy. How and when will the two appear to act independently, and what is the link between them?

TECHNOLOGICAL DETERMINISM

The concept that technology determines the shape of social and political institutions and our relationship with the natural world has a long history. Technology is seen as the force that shapes social change and that determines a whole array of human activities. Closely related to this perspective is the idea that technology is somehow autonomous, that is, that it acts, exists, or affects everything without the involvement of a human agent. Both views have been widely held. Both views reify technology, but this doesn't mean that they are not critical of the consequences certain technologies have had for social life. Many have pointed out, while holding to a kind of technological determinism, that the impact of technology can be deeply corrosive to personal experience, social relationships, and the quality of life. (This view is particularly evident in the work of Marx, Marcuse, and Veblen.) In addition, more recently, many have recognized that technology (industrialization, production, consumerism) destroys nature and that the environmental and ecological costs can be enormous and, perhaps, unbearable. This view has its roots in the works of Thoreau and finds its contemporary expression in Rachel Carson, the Sierra Club, Greenpeace, and—in a more radical vein—Earth First!ers.

In a broad sense, technological determinism appears to be evident in Marx: "The mode of production of material life conditions the general process of social, political and intellectual life" (Marx, 1970). Is Marx a technological determinist? If so, for Marx and for us, technology determines the labor process, the composition of the labor force (who is hired, what skills are in demand, who is kept on, who is let go), the organization of work (what structures and processes are dominant in the work environment, and the social hierarchy of work), and the role that the state plays in determining the structure of production (funding for research, organizing production for war, coordinating production, regulating commerce and, of course, the work environment).

Such a technological determinism suggests that a technological elite will be created to manage this technology, and that technology levels the playing field among the world's nations. In a technologically determined future all nation-states will tend to converge under advanced capitalism. As a consequence, all remaining problems will be technical rather than political in nature. The technological elite will therefore function as a problem-solving managerial class. This view leads to an end of politics, history, and ideological conflict. Choice is irrelevant, human agency falls by the wayside. Power and politics are mystified and obscured.

This perspective makes the problem of choice simple by never revealing that there are choices. Technology determines the direction, and efficiency becomes the goal because that arrangement is rational in a capitalist economy. Efficiency can increase only if the technologies are good. Efficiency equals cost effectiveness. Competition pushes efficiency to the extreme because if you don't use technology to make your operation more efficient, the other guy, business, corporation, or country will do so, and you won't necessarily survive.

> Insofar as the development of technology is the process by which any given task or activity or service is done more efficiently, then technology wins for itself (and those advocating it) political authority and legitimacy. (Street, 1992, p. 27)

Once the idea of an ineluctable link between the virtues of technology and the need for efficiency is established, this becomes the dominant ideological paradigm. Technology is forcing. Efficient production creates a competitive edge, ensures economic growth, and benefits many through a trickle-down effect. Technology thus provides legitimacy and authority.

Therefore, in this view, technology is the key to understanding human social and political relations. For example, inventing the Bessemer converter and process for smelting steel determined the layout for subsequent steel plants. And, at one point, scientists deemed turning asbestos mineral into a "wonder product" that had many uses a "good" idea. In other words, inventions in steel production changed the physical layout of steel plants and the form and nature of steel workers' jobs. Asbestos became a pervasive component of a wide range of products and processes.

Certainly problems arise. Hazards are created, pollution threatens the environment, health and safety threats become evident. But technology can help here too. The expertise (the science, the medical knowledge) to understand the problems can be utilized, and we have the technology (or knowledge) to solve, to substitute alternative solutions until one works. Therefore, the technology continues to determine the shape of things and society in the name of rationality and in the cause of efficiency.

TECHNOLOGY AS POLITICS AND POWER

Another, quite opposite, understanding of technology exists. Simply put, it is that politics, social relations, and power determine technological choices. Technology arises from social choices to fill human needs, or class needs, or male needs. Political power determines those choices. Who has it and how do they use it? Some argue the centrality of the power of the producer (owner of the means of production, individually or as a class); others focus on culture or ideology, emphasizing the domination of male values and rationalist masculinity and the power of ideology in determining the range of choice. Who decides what and how to produce, who decides what the market wants, what the consumer will buy? Are these really choices? The political determinism model says no, we need to understand the social, economic, and political forces that determine the development and use of technologies.

Nevertheless, can we say that all technology is simply the replication of power and control systems present at any particular point in history? Clearly, some forms of technology at certain times and in certain social, cultural, and political contexts appears to have their own internal dynamic. Technologies undergo change and innovation. Workers, consumers, and others who have some power sometimes bring about certain changes. In the social relations model, discussions about technology may be the subject of struggle.

Technology may also be unpredictable in its consequences. No elite, however powerful, can totally control the development and use of particular technologies. Technologies may even occasionally run amok, leaving health and social problems in their wake. Technologies collapse and break down. Some technologies end up not paying off at all—in either profits or improved efficiency. A technology that ends up killing and maiming workers cannot be viewed simply as the heartless acts of a power- and profit-hungry elite. Sick or dead workers and a destroyed or threatened environment are not—within the paradigm—either rational or efficient (the chemical disaster at Bhopal, the nuclear power plant explosion in Chernobyl, and the partial meltdown at Three Mile Island come to mind as examples). Such negative outcomes may even counter the value of technology and thus the presumed motives for using it. The foregoing does not mean, however, that thinking about the probability of disaster or of negative consequences does not occur. It is certainly rational within the paradigm to believe that some risks are inevitable and that some damage will be done to humans or the environment—but that nonetheless *the forecased outcome is worth the anticipated risks.*

The rationality of the technological determinism model finds an echo in the rationality of politics—rational political choices in the pursuit of profit. Street recognizes the complexity of this process:

> The ability to develop a technology is a function of both the way the political process determines what priorities are to operate and the capacity of that system to respond to technical change. *Technology can*

appear to be autonomous, not because it is changing independently, but because the political system fails to control it. (Street, 1992, p. 43; emphasis added)

This is an important point. The link between politics and technology is critical to understanding the development of occupational safety and health problems. The absence of true industrial democracy in the United States and the control over production (of all types) by owners and managers leave workers largely unable to protect themselves.

While the range of social factors that we consider in other chapters (professionals, unions, the public, and functionaries of the state) can mediate and ameliorate the impact of workplace technologies; these social factors mostly act to deal with the problems after the fact. They cannot always do much to anticipate them or—ultimately—to provide a truly healthy and safe workplace.

TECHNOLOGICAL CHOICE

How then does the nature of political control over technology link itself to the political process (in general terms), to the decisions made by government, and to the thousands of individual decisions made by owners, managers, scientists, and engineers? How do the choices get made, and why do they result as they do?

The simple answer, in a capitalist economy, is that technological choices will be made on the basis of efficiency and profitability. A firm will decide to make a certain product in a certain way depending on the market demand for that product. It will choose technologies based on their availability, provenness, alternative arrangements, costs, and a myriad of other micro- and macroeconomic considerations. But, reduced to essentials, price system and the market are what really rule.

In terms of political control, the choice of technology may proceed from quite different imperatives. An owner or manager may choose a certain technology in part because it promises to undermine the influence of skilled workers, largely supplanting their skills while reducing worry over strikes, demands for higher wages, or sabotage in the plant. Certain technologies enable owners to control the time and place of production better, to reduce the costs of labor, and to organize work more efficiently. The assembly line is the classic example. Some technologies are chosen because they result in a monopoly position for the company. Other choices may be made based on demand for continual technological improvement. Embracing a particular technology can inadvertently wed one to related technologies (Zuboff, 1988; Sclove, 1995).

When we say these are political choices, we recognize that they operate at the level of the individual firm, within entire industries, and may be sponsored by the state for political and strategic, economic, or military reasons (technological choices in a wartime economy because particularly critical to economic viability and to military success).

Choices about technology, particularly technology in the workplace, are not made in a vacuum. A whole range of information can be brought to bear to understand the benefits and costs of using a certain kind of technology. The consequences of these choices can also be known (often before they are made). Too frequently those who control production and do not bear the burden of the "externalities" make these choices.

Such choices are biased, sometimes in subtle ways. The dominant ideology and beliefs about efficiency and profitability define management's mindset about technological choice. But the choice is also influenced by the socialization of owners, managers, and the engineers who work for them, and by beliefs, possibly good old-fashioned graft, and institutional structures, as well.

For example, the carcinogenic properties of certain dyes (particularly the aniline dyes) were well understood in Germany, where they were invented and first used in the nineteenth century. Nevertheless, their ability to cause cancer had to be reestablished in the United States after they became widely used here in this century.

Similarly, byssinosis, identified and compensated in Great Britain in 1941, was not "discovered" in the American cotton textile industry until the 1960s. In Hong Kong it remained a "secret" until the 1980s; the first stage of byssinosis, whose symptoms included difficulty in breathing after a weekend away from the plant, went undetected in a country where there was "no weekend" (Wegman et al., 1985).

Technology surely cannot be counted on to protect workers from harm. In addition to our emphasis on involving a human agent in decisions about technology, we also need to stress that those who generally make these decisions rarely make them with perfect knowledge. Few managers will know exactly how a certain technology will operate in a particular workplace, under what conditions it will pose a greater or lesser hazard, what the long-term effects may be, or even if the technology will do what it is supposed to do. Certainly the building contractors and supervisors whose employees worked with or around asbestos did not understand the dire health hazard posed by the substance to workers and the public. Of course, part of the reason us that the Johns-Manville Corporation, the largest asbestos producer, effectively suppressed such information (Castleman, 1979; Brodeur, 1974).

The problem of uncertainty, of course, leads to elaborate structures and methods to control the technology and to understand its problems. The most obvious effort to rationalize this process is risk assessment—the attempt to provide a quantitative or numerical measure of the impact of a certain technology, process, or chemical on human and environmental health. The common justification for such assessment is that it provides an "objective" and "scientific" measure of risk and can lead to reasonable and rational choices. That such a process is deeply political and reliant on a range of inferences from shaky scientific evidence is rarely raised as an issue. The rationality of the process hinges on the acceptance of the objective nature of science and the commonsensical belief that we should weigh costs and benefits clearly before undertaking a course of action (Ginsburg, 1993).

The assessment of technology increasingly relies on the knowledge of specialists. In the work environment, decisions about the health impact of various technologies increasingly rely on the knowledge of doctors, epidemiologists, industrial hygienists, and engineers. As we will argue in Chapter Seven, there are a number of important questions concerning these specialists. How are they trained? Who pays for their research? What are their institutional or class loyalties? What role does "scientific method" have in dominating their perspective? Too frequently the training and affiliations of such specialists can cause them to ignore real human suffering if the causes of that suffering cannot be scientifically linked to exposure in the workplace or the environment.

Risk assessment and the use of scientific information feed the highly litigious procedures by which technology is regulated. Attempts by agencies of the state to regulate technology (the workplace, the environment), often prompted by unions and environmentalists, inevitably result in litigation, as both industry participants and industry associations rarely submit to new regulation without challenging it. Thus, risk assessment usually involves the courts, with all the obvious inequalities in power, information, and access that normally pertain between practiced litigants and comparative amateurs.

TECHNOLOGY AND THE WORKER

Technology (in the broad sense in which we are using it here) has always defined the nature of work. Ever since an early human picked up a rock and used it as a tool, humankind has adapted the material world and shaped it in such a way as to make work easier and more efficient. Since the industrial revolution began, such "technology" has played a progressively greater role in the way human beings make things for survival, exchange, or profit. As tools and machines became increasingly complex and sophisticated, they demanded complementary changes in the structure and organization of the workplace. How those changes were wrought and the impact on work organization reflect the complex relationship between decisions about what technology to use—the shape and form of the technology itself—and the power relationships that emerged between workers and bosses in the industrial age.

As David Noble has argued, the story of modern technology in the United States is the story of the rise of corporate capitalism. The engineers who designed the technology were handmaidens to corporate growth:

> Forces of production and social relations, industry and business, engineering and the price system—the two poles in the dialectic of social production—collapsed together in the consciousness of corporate engineering, under the name of management. (D. Noble, 1979, p. xxiv)

By the turn of the twentieth century, American workplaces were well under the yoke of scientific management. Frederick Winslow Taylor's ideas about how work should be organized and how the worker should be harnessed to technology spread from the newly emerging assembly lines of the factory floor to offices, warehouses, sales floors, and marketing departments. Throughout the United States corporate capital embraced the new management techniques and was to remain wedded to them for most of the twentieth century, but not just in Taylor's most naked form of rationalized production for profit. The first decades of the century also saw the rise of the corporate liberal reform movement. Pushed by a number of related social movements (social reformers, the Progressive reaction to the era of the robber barons, labor agitation by unions, and the political parties of the Left—socialist parties in Germany, Britain, and France, particularly)), corporate capital began to recognize a more effective way of structuring production—one based on Taylor's ideas but also recognizing the value of corporate paternalism in creating a more productive and loyal worker. Resistance by workers and unions to the dehumanizing character of Taylorism coincided with managers' and engineers' dissatisfaction with the narrow scope of Taylor's prescriptions (D. Noble, 1979, pp. 257–284). Thus, scientific management became linked to the schools of human engineering and industrial relations and, by the 1940s it was to blossom into a full-fledged Fordism.[1]

The growth of Fordism in the 1920s and 1930s and the domination of big capital with intensified mass production and mass consumerism resulted in the professionalization and co-optation of health, safety, and emerging environmental movements. In the workplace, the American Federation of Labor spearheaded business unionism, and, as the Great Depression took hold, unions struggled to maintain wages and jobs.

As Fordism gained ground during the war years, it brought with it a vast increase in mass production and the introduction of thousands of new chemicals, processes, and technologies, for most of which the health impacts were unknown. The dynamic of mass production and mass consumption explicitly contained in Fordism successfully hid many of the negative aspects of production (particularly worker health and safety problems and environmental degradation). In the immediate postwar years, with U.S. dominance of the global economy and a quiescent labor movement, little attention was paid to health and safety issues and even less to environ-

---

[1]Here we use the term "Fordism" to denote assembly-line manufacturing using unskilled, mass labor. We also define it (following Gramsci) in terms of a rationalized social and economic system involving not only mass production but also mass consumption.

mental problems, as the growing confrontation with the Soviet Union began to absorb more and more attention.

WORK AND TECHNOLOGY
IN THE GLOBAL ECONOMY

The discussion of the politics of technology illustrates the complexities of the issues surrounding the problem of the work environment (Kuhn & Wooding, 1994a, 1994b). Technology, however, has specific structural effects on economic development, and the rapidity of the change wrought by innovation, automation, and computerization has, indeed, led to a revolution in production. As in the past, the impact has been global. But, today, the rate of change, the speed of communication, and the increasing irrelevance of the nation-state to global economic interaction raise new issues. In what follows we trace the link between rapid technological change, the expansion of the global economy, its impact on national development (particularly in the United States), and the effects these phenomena have had on working conditions.

It has become a commonplace to say that the United States is a "service" economy. The data are compelling: service occupations and industries account for, by far, the largest—and still growing—share of economic activity. In spite of their dominant position within the economy, we know less about service occupations and industries than we know about the now proportionately smaller manufacturing sector. In particular, we are only beginning to form ideas about the effect that this economic transformation is having on jobs and employment conditions.

The increase in service sector employment has resulted in less access to good job benefits for most workers. Fewer employees have decent health care insurance provided by employers, whereas more employees have less paid vacation and fewer opportunities for developing a career path within the company. Most Americans are now working longer hours, for less money (in real terms), than their parents did (Schor, 1991). The increase in stress-related injuries and illness is but one manifestation of the toll this takes on most workers. As new forms of work multiply, new types of dangers become evident—ergonomic problems are increasing, and repetitive motion disorders such as carpal tunnel syndrome are becoming ubiquitous.

Therefore, a full understanding of occupational and environmental disease, the causes of new epidemics (of repetitive motion injuries, for example), the increase in stress-related diseases, and the decline in the availability of good health care for the majority of the population, requires that these problems be understood within the context of the changing structure of work in the United States and the impact of an increasingly global economy.

It is not an exaggeration to say that the United States is in transition to a new era in the production of goods and services. Extraordinary changes have taken place in the past two decades: dramatic increases in international competition, the transition from a creditor to a debtor nation, deep challenges to the sense of hegemony that the United States inevitably asserts from time to time, an extraordinary increase in the rate of capital mobility, and the internationalization of consumer and capital markets. In addition, the breakdown of the "labor accords" of the postwar era and the conservative industrial relations system that accompanied them, the increase in economic inequality within the United States and the erosion of the social welfare system, the widespread experimentation by segments of the business community in the face of unprecedented competitive pressures—all have contributed to the increased economic insecurity of the American worker.

Corporations are responding to intensified competition in a variety of ways, including increased capital mobility and readier resort to contingent or "just-in-time" employment strategies. These approaches have has the effect of externalizing risk to the individual worker and the community. At the same time, the "social safety net" has eroded as a result of the policies of the Reagan and Bush administrations, leaving many people without an adequate level of social services or the help necessary to make the adjustment, as communities and individuals are increasingly made to bear the risks of the marketplace.

The erosion of the middle class, the increasingly unequal distribution of income, the rising incidence of poverty (particularly among children), increased resort to temporary and part-time work, the declining availability of health benefits, and weakened social, environmental, and occupational safety and health regulation—these are all undesirable and potentially destabilizing developments that directly affect public health and welfare.

We believe that to understand public health problems in general, and occupational safety and health in particular, requires an appreciation of what is going on in the American economy and the global economy. We suggest that these developments have added to the health problems facing American workers, especially those unemployed, underemployed, facing job insecurity, working temporarily or part-time, or now part of the ever growing service sector.

More important for the argument being developed here is that these phenomena arise from broad technological change that structurally transforms the lives and working conditions of American workers. Constant changes in work procedures and routines, as well as work organization, reflecting international economic change, are also an aspect of the technological transformations. While the impact of technology on work has been discussed earlier, it is worth recapitulating this process (Bright, 1985; Bell, 1973; Braverman, 1974; Gallie, 1978; Hirschhorn, 1984).

The literature of the 1970s was divided between those who saw technology as upgrading jobs and skill requirements, on the one hand, and those who believed that technology "deskilled" the majority of jobs. The literature of the 1980s and 1990s delivers a consistent message: work organization, skill requirements, autonomy, and other features of the automated workplace are not predetermined. Instead, they are influenced by a host of factors, such as the organization's size and culture, the design of hardware and software, organizational politics, the particular industry in question, and so forth (Karasek & Thorell, 1990; Hartmann, Kraut, & Tilly, 1986; U.S. Congress, Office of Technology Assessment, 1985a).

Effects on jobs include "job enrichment" (in which task variety, skill, and often autonomy are increased), "deskilling" (the removal of these features, creating narrower and more repetitive work), and "job enlargement" (increasing the workload) (U.S. Congress, Office of Technology Assessment, 1985a). Furthermore, the effects on jobs are not static, since the organization and content of work may change significantly before, during, and after the implementation phase of a new technology. Early inefficiencies may give way to increased efficiency, varied and nonroutine work may become tedious as "bugs" are removed from the system, and so on (Greenbaum, Pullman, & Szymanski, 1985).

Even though the effect of technology is indeterminate, technology, as we have noted, is not neutral. The impact technology and automation have in the workplace depends in part on the social organization of work, the amount of control workers have over their working conditions, the availability of alternative employment, and the extent of competitive economic pressures. Increasing use of technology in the rapidly expanding service sector, particularly in clerical and office work, raises a number of significant issues for the health and autonomy of workers.

We have tended to think of the office, for example, as a safe place, by contrast to the dirty, noisy, and physically hazardous industrial shop floor. This perception is generally accurate and is borne out by statistics on the incidence of illness and accidents, which show that industries such as finance, insurance, and real estate, as well as consumer services, have a significantly lower-than-average incidence of illness and accident. Service sector work is far from free of health hazards, however, although the hazards tend to be less visible and dramatic than in manufacturing settings. A number of problems have been associated with office work and several other kinds of nonmanufacturing occupations (e.g., maintenance work, garbage collection), including eyestrain, musculo-skeletal problems, reproductive hazards, other exposures (including exposure to chemicals and other pollutants in the air), noise, poor light, overcrowding, and stress-related illness.

Many jobs now generate a host of ergonomic problems as technology and automation force prolonged periods of sitting and standing in front of computers, check-out scanners, automated systems, and the like. The frequent and repetitive motions required by these jobs strain the musculo-skeletal system and have created a wide range of repetitive motion injuries.

Perhaps the most significant human health consequence of contemporary technological change is the increase in stress felt by most workers across a wide variety of occupations. Stress-related illness is prevalent in the automated workplace, the offices of private and public corporations, small and large retail outlets, and a host of service industries. The state of constantly heightened tension characteristic of those under stress can lead to lower immune function, ulcers, cardiovascular problems, chronic anxiety, and depression. Stress on the job is frequently exacerbated by those organizations that respond to increased competitive pressures by passing this pressure along to their employees. The U.S. Office of Technology Assessment predicted in 1984 that stress-related illness could prove to be the greatest public health problem faced by office workers in the future (National Association of Working Women [NAWW], 1984). The evidence seems to support this prediction.

Autonomy on the job bears an important relationship to stress and health outcomes, since the ability of a worker to control her or his work environment can make the difference between experiencing added workload as a challenge or as a source of negative stress. Control over the pace and methods of work is repeatedly cited as key to worker health, satisfaction, and well-being (Karasek & Theorell, 1990; NAWW, 1984; Hartmann, Kraut, & Tilly, 1986). Lack of work autonomy is particularly common in lower-level service occupations and is characteristic of much part-time, temporary, and contingent work.

In addition to the stress problems resulting from work speedup, longer hours, and increased responsibility, technology used to monitor employee performance adds to the burdens placed on workers. Monitoring of employee performance acts as a constraint on worker autonomy. In the United States the controversy over employer monitoring of workers is framed as a conflict between the right of employers to manage the enterprise, to act to reduce costs, and to avoid potential liability, on the one hand, and employees' rights to privacy and individual freedom (U.S. Congress, Office of Technology Assessment, 1985a). Computerized information technology has increased the speed and scope of potential employee monitoring, expanding the detail and comprehensiveness of information available to management about employee performance and creating new tensions and conflicts between employers and employees.

Such monitoring, perhaps spurred on by competitive pressures that force management to seek myriad methods to increase worker productivity, increases the stress felt by individual workers and exacerbates the tensions deriving from alienation at work and job insecurity.

LESS TIME, MORE WORK

American workers are now working longer and harder than at any time in the postwar period. These extra hours are worked in all industries and by many kinds of workers (Schor, 1991). This expansion of work not only adds to overall stress in the workplace but clearly reduces the amount of time available for relaxation, family, recreation, and leisure pursuits. Longer work hours for those Americans in full-time employment obviously reduces leisure time, as the demands of family life, work, and self-improvement have burgeoned with two-wage-earner families and a decrease in real incomes since the 1970s. Automation and technological sophistication seem to have resulted not, as much of the sociological literature of the 1950s and 1960s predicted, in increased leisure time for workers but rather the reverse (Quinn & Buriatti, 1991).

The restructuring of work has also brought about a restructuring of power in the workplace. The increase in nonunion service sector employment, heightened economic competition, and the changed demographic characteristics of the work force have significantly reduced the ability of workers to resist demands for speeding-up production or monitoring workers' every action, or to fight against the introduction of technologies that create new occupational hazards. In particular, the decline in union power limits the ability of workers to resist pressures for wage concessions and to gain long-term employment security.

The decline of unionization has many causes and would require a long discussion, but one contributing factor has been the changing industrial mix in the U.S. economy. Industries with high levels of unionization are the infrastructural industries (transportation, communications, and public utilities), government, and the goods-producing sector. Service sector industries, the primary growth areas of the economy, have low rates of unionization. Furthermore, part-time workers in all industries also exhibit low rates of unionization, while temporary workers and independent contractors are rarely represented by unions (Appelbaum & Gregory, 1988).

CONCLUSION

In the changing American workplace much is made of leaner, meaner production. In many workplaces, management has achieved remarkable production gains and improved the bottom line. That this was possible in the 1980s had much to do with new technologies, increased automation, and capital flight to countries where labor is cheap and government regulation weak. However, it also reflects a changing set of beliefs about how management's (and the state's) responsibilities to workers are defined. Stressing the virtues of trickle-down economics and emphasizing the need for capital to be free from governmental—that is to say, social—constraints (environmental and health and safety regulation in particular, taxation in general) created a new ethic of personal responsibility and reaffirmed the seeming virtues of the market. What little sense of obligation there had been of employer to employee (to provide relatively secure employment over the long term, to provide steadily increasing income and at least an elemental set of benefits) has broken down as the structural pressures for change are played out in the form of ideological assertions of the rights of capital over the rights of workers and citizens. Attacks on governmental regulation and organized labor, often spearheaded by business, have exacerbated job instability and insecurity. There is no longer a commitment by the government to full employment.

Clearly the changing character of the U.S. economy has altered the nature and experience of work for many Americans. It also has enormous social and political implications. The decrease in heavy manufacturing, the expansion of the service sector, and the concomitant changes in benefits and unionization have all contributed to changes in the overall welfare of working people. But the structural changes we have described in this chapter have had a profound impact on the nature of employer–employee relations that goes well beyond the restructuring of work and the availability and extent of benefits such as health coverage and vacation time.

What we argue here is that these changes have resulted in some new clearly identifiable problems that shape the overall welfare of Americans and have important health consequences in and outside of the workplace. In broad terms, we suggest that technological and economic changes place an increasing burden on workers, and that there are now new chemical and ergonomic hazards facing workers at their workplaces. Further, working now takes up a greater proportion of time than in recent memory for those in full-time work, resulting in a significant reduction of leisure time and a negative impact on the quality of life for many working Americans.

We relate these problems not only to the issue of the restructuring of work as a response to global economic changes of the 1970s and 1980s but also to the decreasing ability of many workers to resist these changes. This decreased ability results from the decline in trade union power in the workplace and the dominance of an ideology that promotes the rights of corporations and limits the acceptability and legitimacy of state interventions to protect worker and environmental health.

CHAPTER FOUR

The Social
and Political Context
of the Work Environment

Work is a necessary human activity. People work to survive. Yet, work is more than simply drawing a paycheck. Work provides a host of rewards—and problems. It may be laborious and numbing—or stimulating and satisfying—frustrating and demeaning. All too often it is dangerous and unhealthy.

Work occupies a central place in most people's lives, but the overall context of work is often ignored or poorly understood. To recognize and prevent work-related disease and injury requires that health care providers and other health and safety professionals appreciate the full context of work and workplaces in the world today.

The magnitude and patterns of occupational disease and injury in a particular society are strongly affected by the level of economic and technological development, the societal distribution of power, and the dominant ideology of a particular social and political system. All of these factors bear on the way in which diseases and injuries are "produced," recognized, and prevented, and the extent to which workers receive compensation for them.

IDEOLOGY

The organization of work and the role played by key actors in the workplace are deeply influenced by ideology—the set of beliefs, norms, and values of workers and managers, bureaucrats, scientists, and others that reflect what they think about society and about themselves, and what they think they can expect from work, employers, government, and one another.

A capitalist, free market economic system incorporates presumptions about human behavior that most people have come to accept: ideas about individual "choice" and "rights," belief in the primacy of private property and the efficiency of markets. Liberalism asserts that government should not constrain these rights without good reason, and Americans, in particular, are deeply suspicious of government. It is, therefore, necessary to examine the role of ideology in order to identify the assumptions that determine power relations in the workplace and how they are reflected in the problem of occupational health and safety.

The typical workplace in the United States is structured hierarchically, with the owner or owners at the top and managers, supervisors, and workers. in successively subordinate positions. This hierarchy of command reflects the distribution of power in an organization. At work, this means that the owners and managers have complete control over investment decisions, the company budget, what is produced, and how and when production occurs. It also means nearly total control over the hiring and firing of workers and, ultimately, nearly total control over the conditions of work.

Unions are considered to be a potential counterweight to this one-sided power arrangement. Where labor unions exist, they have had some success in winning better wages and working conditions from employers. The achievements of labor unions, however, have usually been tempered by several factors: the strength of the economy, the level of unemployment, the economic and political strength of organized labor itself, and an ideology that supports the rights of property. The persistency of labor's achievements has also depended importantly on the extent of government support for protecting and promoting the rights of workers.

In Europe, although the private property rights are held in high regard, the power of unions and workers' parties and the widespread acceptance and expectation of government regulation of working conditions have resulted in a greater ability by government to regulate private industry and working conditions, as compared to the situation in the United States.

The culture of most liberal democracies, including that of the United States, has provided a uniquely fertile ground for belief in the rationality and apolitical nature of science and technology. The social and political systems in these countries reflect a continued belief that social and public health problems—indeed, most societal problems—are amenable to technical solutions. Remarkably enduring has been the ideology of the "technical fix." It is the separation of science from politics and professionalism, that is, from issues of power and control.

These ideological assumptions are part of a set of powerful beliefs that impinges on the work environment. Scientists, engineers, and industrial hygienists are all taught that their expertise (and the technological solutions it represents) can mitigate workplace hazards. While this is certainly true, it ignores the social and political context in which workplace hazards arise. This context defines how the workplace is designed and managed.

MANAGEMENT THEORY
AND THE STRUCTURE OF WORK

Although under attack and reconsideration in recent years, the general tendency in management theory from the time of Adam Smith, the father of economic liberalism, to the present has been to divide work into ever more discrete units to increase productivity, cheapen the cost of labor, and increase management's control over the labor process. This quest for "efficiency" became more self-conscious and explicit in the early twentieth century with the work of such promoters of "scientific management" as Frederick Winslow Taylor (Braverman, 1974; Buroway, 1979). In Taylor's view, the worker should be treated not as a whole person but rather as a collection of machine-like movements: walk, bend, grasp, sit, depress typewriter key, and so on. Such motions can be analyzed, timed, and reassembled into a program for maximum productivity. This "scientific" approach to management was widely accepted, both in capitalist and noncapitalist economies.

Taylorism had wide-ranging implications for the quality of work life. It meant the separation of conception from performance and the division of performance into multiple repetitive tasks. The intrinsic satisfaction of "work"—craftsmanship and the ability to take pride in the whole finished product—has necessarily been diminished as a consequence. Employers have increasingly relied on supervisory hierarchies and monetary rewards and punishment, such as piece rates and bonuses, to motivate workers in a carrot-and-stick fashion. While Taylorism was supplanted by Fordist policies during World War II and came to dominate work organization in the postwar years, the American workplace remains characteristically hierarchical. The discipline of capitalist work has become more evident where well-paying jobs are scarce, immigrant labor is widely used, and unions largely absent.

Another profound influence on modern production and the workplace has been the rapid increase in the use of chemicals, especially since the end of the war. There are currently 70,000 chemicals in use in the United States, with 1,000 new chemicals introduced each year. A similar number of chemicals and chemical processes are evident in most of the industrialized world, and increasingly so in developing countries as production is shifted to them. The vast majority of these chemicals are unregulated, and their human health effects are unknown. They are used in a variety of work settings to manufacture or process a wide range of products, but they are also encountered in a range of occupations not traditionally considered dangerous. From typists and stockroom workers to janitors and artists, workers confront some potentially toxic chemicals on a daily basis. This revolution in materials is distinct from the impact of "scientific management" (discussed earlier), but, because so little is known about the potential health effects of these chemicals, it has had a much more profound effect on the day-to-day experience of workers than one might expect.

Further, technology has increased the speed of production enormously—putting yet greater pressure on workers to perform rapid and repetitive motions that are damaging to mental and physical health. Stress and related psychological and physiological illnesses are increasing in industrialized countries, including the United States, as the pace of work and life, as well as pressures to work longer hours in order to compensate for declining real wage rates and standards of living, increases. Some countries, however, particularly those in postwar Europe, have shortened their workweek, owing largely to historical and cultural factors as well as pressure from powerful trade unions. More recently, trends toward higher unemployment have ratcheted up the call for an even shorter workweek (Schor, 1991).

With the speedup in production has come considerable automation. Apart from the fairly obvious potential physical hazards associated with use of robots, robotic systems, and of highly automated machinery, there is also the way in which automation takes away jobs, deskills others, and leaves fewer workers in charge of complex systems. Today, in many plants, with the help of automation, one worker does a job that may have required ten before. That person more than likely now faces enormous stresses and responsibilities and may even be required to work punishing hours. Under these circumstances, the promise of automation to replace grueling mindless labor has resulted in more stress, longer hours, and overwhelming responsibility at work for the relatively few workers who remain in place (Zuboff, 1988).

Economic and technological changes go together. The spread of new technologies, the globalization of the world economy, and vast changes in the international division of labor both directly and indirectly affect not only the work environment but also power relationships generally in society.

THE DISTRIBUTION OF POWER

All societies are composed of classes, of interest groups, of sects and sectarians, of minorities and majorities, with varying degrees of power and influence. The distribution of such political and economic power and influence is another essential factor shaping the work environment. In the most simple formulation, there are "workers" and "owners." In advanced industrial societies, such a formulation cannot capture the complex features of the contemporary class system. In such societies, a middle stratum has developed (composed mostly of independent professionals, an enduring class of small business owners, and a growing group of government employees) with a wide range of social functions and with their own roles, interests, and power. The varying degrees of political power among lower, middle, and upper classes sets limits on what can happen in a particular workplace or a particular industry. Class and the distribution of power in society, therefore, shape the work environment.

Thus, in addition to the development of the market, the level of technology, ideological considerations, and changes in the global economy, power distribution related to class, race, and gender constitutes the framework in which the actors in an industrial system attempt to create a "web of rules" governing the work environment. Management, labor, and government are constrained in their behavior by these broad social–environmental factors.

In many liberal democracies, although the stated goal is equality, power in society at large is unevenly distributed along the lines of class, race, and gender. Because the workplace is a microcosm of society, the power relationships in society are reproduced at work. Inequities in the distribution of power have a profound influence on work and health because power determines who does what work and under what conditions, that is, who gets exposed to risks and what is considered an acceptable risk. Furthermore, the people affected are not the people deciding the acceptability of workplace hazards:

> All the clichés and pleasant notions of how the old class divisions . . . have disappeared are exposed as hollow phrases by the simple fact that American workers must accept serious injury and even death as a part of their daily reality while the middle class does not. Imagine . . . the universal outcry that would occur if every year several corporate headquarters routinely collapsed like mines, crushing sixty or seventy executives. Or suppose that all the banks were filled with an invisible noxious dust that constantly produced cancer in the managers, clerks and tellers. Finally, try to imagine the horror . . . if thousands of university professors were deafened every year or lost fingers, hands, sometimes eyes, while on their jobs. (Fussell, 1992)

Social class and class-based assumptions have been widely discussed from a variety of perspectives: sociological, economic, and political. Class is clearly related to family background, level of education, occupation, and a variety of cultural factors. The lower the social class of an individual, the less likely that he or she will be to have range of educational and employment options. Class frequently determines levels of material well-being and health. Because class influences employment options, it affects the probability of becoming ill or injured at work.

THE IMPACT OF RACISM

In the workplace and in society as a whole, racism plays a role in determining who does what job, how much he or she will be paid for it, and what alternatives are open. For most of its history, the United States has depended on minorities to do the least desirable and most dangerous work. Immigrant and minority communities have been the major sources of labor to build the railways, pick cotton and weave it in the mills, work in the foundries in the automobile industry, run coke oven operations in the steel industry, sew in the sweatshops of New York's Lower East Side, and provide migrant agricultural labor. Minorities are still overrepresented in the most hazardous and least desirable occupations.

In addition, minority workers may leave a hazardous work environment only to arrive home to a hazardous community environment. Since the early 1980s, in the United States, scientific evidence has increasingly pointed to discriminatory environmental practices of certain industries, state and local governments, and in some instances the federal government. The overall assessment is that minority communities are home to a disproportionate number of toxic threats to their health (Bullard, 1992).

A social system with strongly racist overtones effectively bars members of minority groups from holding significant positions of power and, consequently, elevates the concerns of dominant racial or national groups. For example, one of the key reasons for the lack of attention to hazards faced by farm laborers in the United States (most of whom are African American or Latino) is their relative lack of power in the American political system.

In the United States, racism as a political phenomenon is of particular significance because of the history of European Americans' conquest of Native American and Latino peoples and the slavery of African Americans. This bitter history has great importance for contemporary politics and, therefore, for the work environment and its regulation.

THE IMPACT OF SEXISM

Any discussion of power relations must include the situation of women, whose experience of work is generally different from that of men. Most obviously this disparity is reflected in the wage differential between men and women for comparable work. Despite a political and legal commitment to equality, as of 1991, women were earning only 70% of men's pay for the same work, and the gap widened the higher one ascended the career ladder (Lewis, 1992). African American women and Latinas earn only 50 percent of white men's pay (Amott, 1993).

Even though women frequently work outside the home for as many hours as their spouses, domestic duties are rarely shared equally. Working mothers sleep less, get sick more, and have less leisure time than their husbands. One study found that women who were employed full-time outside the home and whose youngest child was less than five years old spent an average of forty-seven hours per week on household work, while their male counterparts spent only ten hours on household tasks (Byant, Zick, & Kim, 1992). The average working woman put in an estimated 80 hours a week in both job and household work and up to 105 hours if she had sole responsibility for her children. Thus, inevitably, stress and fatigue resulting from the demands of having to balance work and home life remain serious problems.

Women are also the main targets of sexual harassment at work. Any unwanted verbal or physical sexual advance constitutes harassment, and this can range from sexual comments and suggestions, to pressure for sexual favors accompanied by threats concerning one's job, to physical assault, including rape. Studies indicate that 40 to 60 percent of women have experienced some form of sexual harassment at work. An estimated one-third of the largest 500 companies in the United States spent approximately $6.7 million in dealing with sexual harassment in 1986 (Spangler, 1992). That sum is considerably larger today.

Gender relations have political—and hence work environment—implications. Cultural assumptions about gender can have a strong impact on the distribution of power in society. A strongly patriarchal society that bars women from positions of power is also likely to have a profoundly sex-segregated labor market. As a result, sexual harassment and occupational health in female-dominated retail trade jobs may not be considered as important as in a white-collar environment.

THE MICROCONTEXT
OF OCCUPATIONAL HEALTH:
LABOR–MANAGEMENT RELATIONS

Work and Workers

In addition to suffering the detrimental effects of lack of control, which by itself causes stress, workers find that their interests typically conflict with those of management. Management's goal is to maximize profits, labor's goal is a fair wage for a fair day's work. Expenditures on health and safety are normally seen by management as limiting profits. As a business school textbook advises:

> In making decisions about their workplace, managers have two choices. They can remedy health and safety problems or they can provide risk compensation to workers. If reducing risk is less costly than the additional compensation, then working conditions will be improved. However, if the marginal cost of worker compensation is less than the marginal cost of safety improvements, then the firm will choose the compensation alternative. This outcome represents an efficient allocation of resources in that the firm minimizes its total costs. (Peterson, 1989, pp. 429–430)

Although one might hope that managers' exercise of their conscience might offset their training and the incentive system within their businesses, history has shown that it is unwise for workers to depend on the sheer benevolence of management. The sociopolitical setting provides only weak motivation for management to construct a safe workplace. Government regulations exist, but they are not always enforced (which should not be so surprising, given that there are five times as many fish and wildlife inspectors in the United States as OSHA inspectors).

Labor–management relations may be particularly problematic when the potential loss of jobs is pitted against improving occupational or environmental conditions. The most frequent example of this dilemma occurs under conditions of "job blackmail"—a colloquial term for the problem created when workers are forced to choose between remaining in a hazardous job or finding employment elsewhere (Kazis & Grossman, 1982). Examples include employers who threaten to fire workers or relocate the plant if workers or regulatory agencies try to impose controls over hazardous production. Job blackmail is found more often in those workplaces where workers have little or no control over their jobs as well as in those workplaces that are not unionized. Although not unique to minority workers, job blackmail takes an especially heavy toll on them, because they are more likely than nonminority workers to hold hazardous jobs. Although job blackmail may occur in a variety of direct and indirect ways, the end result is to force workers to choose between being employed or not.

In job blackmail the choices are seldom, if ever, favorable to the worker. The worker who chooses to remain on a hazardous job may, in the short term, avoids unemployment, but may seriously jeopardize his or her future health and safety. The worker who chooses not to question "unfair" compensation will continue to receive a paycheck but will still earn less than she or he is worth. The worker who chooses not to unionize may remain employed, but will likely remain employed in an unjust, unsafe, and unhealthful workplace. Even in those situations where a worker remains on the job, she or he may be labeled a trouble-maker and ostracized to the point of having to quit the job anyway.

While unions are a force in spurring companies to attend to health and safety problems—typically through collective bargaining agreements or, where they exist, union-controlled health and safety committees—and government regulation provides a further stimulus, there are two other motivational sources for employers to improve health and safety: (1) corporate reputation ("public relations"), which functions to press management to not appear negligent in their provisions for workplace safety (although it tends to function more effectively for pollution problems and environmental concerns than for work-related illness and more often pertains to large rather than small corporations); and (2) the cost of replacing labor. If a company has invested in developing a skilled and loyal work force, it is unlikely to want to damage that investment by exposing workers to dangerous conditions. This factor helps to explain why low-skilled, easily replaced workers, such as migrant laborers or poultry workers, are so vulnerable to managers' single-minded pursuit of profits wherever such mutual loyalties are absent.

The Changing Structure of Work

As we noted in Chapter Three, the economies of the United States and many other developed countries are changing rapidly. The shift out of heavy manufacturing and toward the service sector affects the structure of work and the work experience for many Americans. In general, in service industries, the most rapidly growing sector of the economy, wages are lower, benefits minimal, job security limited, and unions virtually nonexistent; moreover, much of the work is part-time or temporary.

In response to the shrinking economic pie of the 1980s and 1990s, employers are increasingly using part-time and temporary workers to cut costs. The average part-time worker earns only 60 percent of the hourly wage of a full-time worker. Fewer than 25 percent of part-timers have employer-paid health insurance, compared to nearly 80 percent of full-time workers. Sixty percent of full-time workers have pensions provided by employers, while only 20 percent have this coverage (Amott, 1993). In the United States, as of 1990, there were some 5 million *involuntary* part-time workers—that is, workers who would prefer to be working full-time but were unable to do so, given the job environment.

In addition to lower pay and fewer benefits, this trend toward temporary work and part-time work has other negative aspects. Temporary workers live with the stress of not knowing when and for how long they will work. They have little or no job security. Neither part-time or temporary workers receive equal protection under government laws, including occupational safety and health regulations, unemployment insurance, and pension regulations. Few are represented by unions (Amott, 1993). A case study commissioned by OSHA of contract labor (usually small companies of nonunion workers, brought into a plant to do maintenance and other work) in the petrochemical industry shows that contract workers get less health and safety training and have higher injury rates than noncontract workers (John Gray Institute, 1991). The long-term consequences for occupational health and safety of an increasingly unorganized, temporary, and part-time work force should not be underestimated or dismissed lightly.

Another increasingly common characteristic of changes in the structure of work in the United States is the rise in home-based industry. In 1949 Congress passed a law making industrial homework illegal, largely because it was almost impossible to enforce workplace regulations and labor standards (such as the minimum wage) for homework. The Reagan administration pushed Congress to legalize homework, resulting in the rapid growth of home-based manufacturing and service work throughout the 1980s. A large number of homeworkers are women, typically garment workers and clerical workers who are paid on a piece-rate system. Piece-rate payments place a premium on speed, increasing the risk of accidents and exacerbated repetitive motion injuries and thus resulting in numerous ergonomic problems (in part reflecting the fact that the home is not likely to be properly equipped for the type of work being undertaken). Chemical exposures also pose a problem. Semiconductor manufacture undertaken at home, for example, not only exposes workers to toxics used in the manufacturing process but also may contaminate local sewerage systems.

Work and Labor

Of the 110 million Americans in the work force, approximately 92 percent work as employees of others. Approximately one-fourth have professional, managerial, or supervisory employment, with varying degrees of partial autonomy and control over their own jobs. These people both work and labor. Most workers, though, only *labor*. They find what jobs they can and, by and large, do what they need to do to keep them. They do not choose what they will make, under what conditions they will make it, or what will happen to it afterward. These choices are made for them by their employers, the sales and labor markets, and the workings of the economy as a whole. Whatever control most workers have over how much they receive in return for their labor, how long they labor, how hard they labor, and the quality of the workplace environment is acquired in a contractual situation in which the workers' desire for comfort, income, safety, and leisure is continually counterbalanced by the employers' need for profit.

Many workers have profound ambivalence about their jobs. While labor provides an income, workers also seek less tangible satisfactions from their work. What workers love about work is the opportunity to guide their own lives, to do meaningful things, or "to get back from the activity not only the physical means to live, but also a confirmation of significance, of the process of being oneself and alive in this unique way" (Williams, 1968). Conversely, what workers most oppose is when their work is made seemingly meaningless by management's championing of efficiency, productivity, and profit at the expense of integrity, autonomy, and creativity.

Modern production and market competition lead employers to seek the highest possible rates of productivity. The normal social interactions among workers, which in a less mechanized and fragmented work process appear as part of the rhythm of work itself, are seen as disruptive of production, and attempts on the part of workers to establish some level of control and sociability in the workplace are often misconstrued. Employers and managers who see such acts as threats to productivity and efficiency consider them to be indications of laziness. Workers, even in nonunion settings, may view them in a similar way or as efforts to protect themselves against the requirements of a fragmented division of labor that treats them as tools rather than as people. These attempts at greater personal control actually represent, consciously or unconsciously, the individual's desire to replace labor with work. But the structure of the contemporary workplace undercuts such acts of nonconformance or self-assertion.

Contemporary innovations like word-processing technology, computerized record-keeping, electronic mail networks, and computer and video monitoring have turned large offices into disaggregated assembly lines. New forms of work organization have broken down the close personal tie that often used to exist between secretaries and their bosses, and new technology has downgraded the skills required. With these changes, clerical work becomes subject to the same kind of machine-like analysis and control as factory work.

Similar situations are often found in service, retail, distributive, and other types of work. What is true for the auto worker, the word processor, and the keypunch operator is increasingly the case for the short-order cook, the checkout clerk, and the telephone operator.

An increasingly fractionated division of labor and "scientific" work discipline are ways of exerting managerial control in the interests of efficiency and profit. But the experience of alienation and powerlessness on the part of the workers is not limited to workplaces where this type of organization is imposed. Many jobs in small shops—particularly in the service and retail sectors, which employ the largest number of women and youth—are equally unattractive despite their lack of specialization, and frequently they are dangerous.

The characteristic jobs of a service sector economy tend, therefore, to replicate quite often the alienating, repetitive work once associated with assembly-line production and the monotony of the modern factory. Today, in developed countries, however, improved technology and the ubiquitousness of computers have enormously increased the potential pace of work as well as the ability of the work rate to be monitored. Technology combines with pressures for increased productivity in an increasingly competitive world economy. "Competitiveness" and the drive for productivity entail enormous long-term costs to the health and well-being of workers, an observation not often made by the media.

The constant demand to do work faster and more efficiently, to "produce" under the whip of being fired or laid off, takes an enormous toll on the mental and physical health of workers. The dignity of work is not evident in the voices quoted above. Whenever American competitiveness in foreign markets declines, cheaper goods and services enter the United States from developing countries, where workers get wages little above subsistence and often work in horrendous conditions; as a consequence, there is even greater pressure on domestic manufacturing to "compete." The reality of that competition for most American workers ends up taking the form of wage give-backs, compulsory overtime, work speedup, and less and less attention to workplace health and safety.

ORGANIZED LABOR

Unionization is a way of counteracting the disempowering, disenfranchising effects of class, race, and gender. Unions provide workers a voice in determining the rules and conditions of work, wage rates, and benefits. They have collective strength that counterbalances management's power and prerogatives. Some unions have been deeply involved in health and safety issues, but, for most unions, health and safety issues are only ancillary to more important concerns. In the United States, given the weakness of unions and the historic antagonism to labor, unions have not always been able to dedicate the necessary resources to protect their members from workplace hazards. In Europe, organized labor has been more successful in combating the prerogatives of management, and, in a number of European countries, social-democratic political parties supported by labor movements have frequently come to power. Even in the United States, with its relatively weak labor movement and the absence of social-democratic or labor parties, unions do offer protection against arbitrary exercise of authority and do provide for injured workers.

Formally, unionized workers try to regain some control over the labor process through collective bargaining—the negotiation of work rules and grievance mechanisms, the institutionalized process for adjudicating individual complaints. However, only approximately 15 percent of workers in the United States are unionized, and even where grievance mechanisms exist, the mechanisms are not always utilized. Informally, workers seek what escapes they can find or fabricate. They sneak a surreptitious cigarette, they fantasize, they horse around, and they fight. "Anything so that you don't feel like a machine" is a common refrain.

Organized labor in the United States is now weaker numerically and politically than at any time since World War II. This decline began in the 1970s and has worsened throughout the 1980s and 1990s. The decline is evident across the whole range of union activity: loss of negotiating strength, decrease in membership, decline in strike activity, and a vast increase in "concessionary" collective bargaining agreements between unions and industry.

The strength of a labor movement determines a host of issues that directly influence worker health, including what information is generated about workplace hazards, who has access to it, what workplace standards are set and by whom enforced, the options open to workers encountering a hazard, and the effectiveness of workers' compensation (Elling, 1986).

Unionized workers are more likely to be informed about the presence of health and safety hazards than nonunion members in the same jobs (Weil, 1992). In addition to union-sponsored educational programs, the union provides a shield against employer discrimination. This shield is extremely important for health and safety because employers may fire a worker for raising concerns about health and safety problems.

Unions in the United States and elsewhere have sought to create legislation requiring employers to clean up the workplace, to enforce employment regulations concerning women and children, to abide by proper hours of work, and to set and enforce industrial hygiene standards. In the United States, where OSHA requires that workers be informed about the hazards associated with the chemicals they work with, unions have pushed to make sure that these "right-to-know" regulations are complied with. When there was no federal right-to know law, some unions negotiated this right, as well as the right to refuse unusually hazardous work.

Unemployment

Unemployment is more destructive to physical and mental health than all but the most dangerous jobs. Recent studies have even suggested a correlation between high unemployment and high mortality from heart disease, liver disease, suicide, and other stress-related ailments. Interestingly, changing levels of unemployment have an impact not just among unemployed workers themselves but also within their families. For example, households in which the husband is unemployed or underemployed show rates of domestic violence two to three times greater than the households of fully employed men. One of the most striking research findings on unemployed workers is the near unanimity with which workers internalize the experience of joblessness as one of personal lack of worth. This sense of worthlessness appears completely unrelated to a worker's actual degree of responsibility in losing his or her job.

In the contemporary American economy, the unemployed, like the poor, are always with us. In the 1980s, the unemployment rate in the United States fluctuated between 6 percent and 11 percent (some economists have even suggested that a 5 percent unemployment rate should be considered "full employment"). In the 1990s the current expansion has resulted in the lowest levels of unemployment in recent decades. But the burden of unemployment falls most heavily on minorities, whose unemployment rate (particularly among young black males) is up to three times that of the national level.

Even more than in the United States, unemployment has become a severe problem both regionally and internationally. Among the member nations of the European Union for example, the official unemployment rate was 11.6 percent in 1997 (Economic Indicators, 1999). Moreover, this statistical focus is only on those actively seeking work; by excluding those jobless who, through discouragement, have stopped looking or never began, the official data understate the magnitude of the problem. Such figures rarely include the underemployed, those working part-time who seek full-time work, and those women who would be working if attainable and attractive jobs were available. In the developing world, the percentage of workers who are unemployed or underemployed is often much higher than in the United States or Europe.

Unemployment has significant economic effects. The existence of so many unemployed people keeps wages down as more people compete for jobs and are willing to take lower wages in the struggle to earn a living. It also makes union organizing extremely difficult. As workers lose jobs in traditional manufacturing—where unions had strength—and as management campaigns against unions, the barriers to encouraging people to join unions or to mount significant organizing drives become larger. All these factors weaken the movements to protect workers from occupational hazards.

The Role of Management

Unquestionably, there are capitalist firms that seek to maintain safe and healthy work environments. These are frequently large, profitable companies that have relatively secure markets for their products and that have decided that their continued success depends on a well-motivated, high-quality and healthy work force. Frequently these are firms that have a deep commitment to collective bargaining and to negotiating industrial peace. Some firms have decided that the only way that they can attract and keep highly skilled workers is to ensure the quality of working life. And other firms, concerned about product safety because of consumer concerns or the inherent risks of their technology, have attended to worker health and safety virtually as a spillover from their other essential activities.

The remarkable success of Japanese industry in reducing its injury rate—probably as a consequence of its attention to quality in general and its abhorrence of waste—may have beneficial consequences in U.S. and European firms' pursuing Japanese-style manufacturing success. Sometimes these company efforts may miss problems associated with low-level chemical exposures because they focus primarily on the more obvious safety hazards; nevertheless, such efforts are to be applauded.

Small firms, too, can pay serious attention to safety and health hazards, particularly when the owner or manager came up from the ranks, knows the processes well, and maintains close social contact with the employees. But the economic pressures on small companies may undercut even the most decent employer/master craftsman. For small or large firms, the pressures of the market are hard to resist. In these cases, the role of government in enforcing work environment standards is particularly important.

CONCLUSION

In this chapter we have argued that the work environment is a microcosm of the ideological, social, and political relations of the wider society. In the United States, this means that beliefs about individual rights, the rights of property, the role of government, and established power relations are often played out in the workplace—that is, that the social and political organization of work reflects the wider society.

Understanding the impact of workplace hazards and the systems established to ameliorate them should take into account how the political and social context determines the incidence, recognition, definition, and prevention of occupational disease and injury.

CHAPTER FIVE

The Politics of Regulation

In 1997 farm worker advocacy groups in Florida examined the state-provided regulatory protections for migrant workers. They found the following problems:

- The State repeatedly failed to find a causal connection between pesticide exposure and the injuries suffered by farm workers. In only two instances did it conclude that pesticide exposure led to worker injury.
- The State found regulatory violations in thirty-one instances, but issued only two fines.
- The State failed to adequately investigate poisoning complaints even when a farmworker was seriously injured or killed, by systematically: failing to interview coworkers or other eyewitnesses out of the presence of supervisory personnel (with adequate translators); failing to obtain relevant medical records; routinely accepting uncorroborated employer claims of compliance; using checklists as a substitute for a thorough on-site inspection; and ignoring evidence of employer retaliation.
- The State lacked adequate investigative protocols: it routinely failed to collect soil, plant, clothing, and other physical samples that would have enabled it to verify exposures and identify the pesticide(s) used; it routinely failed to draw reasonable inferences from the information obtained; and it failed to make regulatory determinations based on objective, corroborated evidence.
- The State failed to coordinate the investigative efforts of the

Florida Department of Agriculture and Consumer Services (FDACS) with those of other enforcement agencies, such as OSHA, or the Division of Workers Compensation. It also failed to ensure that FDACS established effective communication with health providers who are required by state law to report pesticide exposure incidents.

• The State failed to impose meaningful penalties when pesticide violations resulted in worker injury. Moreover, Florida went to great lengths to avoid the conclusion that pesticide exposure led to worker injury. For example:

• In two incidents, FDACS noted both that unprotected workers had been placed in a field before the REI [Restricted Entry Interval] expired and that the workers subsequently received medical treatment. Nonetheless, it refused to find any connection between the exposure and the subsequent injuries.

• In another incident, FDACS found that an employer unlawfully caused a pesticide to drift onto an area where farm workers were working, but failed to find any relationship between exposure and subsequent worker injury.

• In another case, the Department noted that a farm worker was accidentally sprayed with pesticides and that the company had failed to provide him with prompt transportation to a medical facility. However, it drew no conclusions concerning the relationship between the exposure and the worker's injury.

The report concluded by noting:

> By failing to adequately investigate complaints, issue citations for pesticide poisoning and impose meaningful penalties for serious WPS [Worker Protection Standard] violations, the State has deprived farmworkers of adequate protection and wholly undermined its effort to deter future misconduct, as the repeat violator complaints in FDACS' files amply demonstrate. (Davis & Schleifer, 1997)

The conditions of the work environment are determined, as we have noted, by a variety of factors: the level of technological development, the social organization of work, the structure of economic development and the balance of power between workers and managers both within and outside the workplace. Regulatory policy emerges from the intersection of these social forces or factors. One key issue, however, is in what ways the conditions of work are determined by regulatory policy. In this chapter we provide an overview of the development of regulatory policy in the United States and begin to explore how various actors related within the work environment, paying particular attention to the role of the state.

THE WORK ENVIRONMENT AND THE POLITICS OF REGULATION

The relationship between the institutions of state control and the social and economic forces that govern safety and health at work provide an opportunity to better understand the political economy of the work environment.

It might be asserted that state control of health and safety arises from the needs of the capitalist system as a whole. The necessity of ensuring the reproduction of labor power and by extension ensuring the conditions for the accumultion of capital are manifested in government legislation to protect people at work. The specific levels of capitalist development determine the particular form of this intervention.

The problem is that defining regulation as determined by the structure of the economic system, tough it may suggest possible constraints and imperatives on employers, takes no account of the potential of class or political struggle in either prompting intervention or in determining the form it will take. It also says very little about the way various institutional structures mediate between employers and workers by determining the types of allowable struggle (e.g., the role of parliamentary processes and the actions of political parties, the nature of industrial conflict and the legal constraints on strike action, the existence of general legal frameworks and the ideological power of democratic norms and values, etc.) or the way in which institutions structure policies (e.g., the passage of laws such as the Occupational Safety and Health Act [OSHAct] and the creation of administrative and enforcement agencies that follow a particular pattern of state intervention).

State theory rarely considers the role played by the conscious acts of individuals or collectivities in determining either the *forms* of the state that have emerged (the history of pressures for democratic political arrangements, the impact of wider class struggle on state institutions), or the *way* in which various struggles circumscribe the possible kinds of interventionism (the class conflict model).

For example, the long history of factory legislation in Britain reflects the development of the modern administrative state from the Victorian era to the present, and this, itself, is a consequence of extensive political and economic pressures to reform laissez-faire capitalism. The creation of the Factory Inspectorate was the result of political and social pressures emanating from a newly formed and organized working class. This development in turn created institutional arrangements and a body of legislation against and within which further political struggles over workplace health and safety have been played out.

Although the class conflict model emphasizes the importance of class struggle in determining the development and shape of state intervention, it, too, has some fundamental problems in explaining the existence and shape of health and safety regulation. While it might assume that class struggle creates pressures for reform legislation and that the history of health and safety in the United States reflects a considerable amount of worker and trade union pressure for state intervention, it does not explain either the particular forms of this pressure (strikes, demands articulated through representative political parties) or the forms of the response (type of policy, nature of legal or administrative organization).

In addition, the class conflict model tends to assume an instrumentalist view of the state, wherein the state is "up for grabs" and therefore the working class is forced to compete for control of the state, wrenching control away from the dominant class. As we have seen, this argument assumes a great deal of class homogeneity and a degree of class consciousness (the extent to which a class is a "class-for-itself") that contradicts the historical reality of class organization in the United States.

Further, a simple class conflict model cannot entirely explain the extent to which the state may operate relatively autonomously. This shortcoming is especially pertinent in the case of the work environment, where state intervention has meant considerable economic and political costs for private industry and where the imposition of such regulation has been virulently attacked and resisted by the business community.

Health and Safety Regulation in the United States

The earliest factory legislation in the United States, as in Britain, was developed around the textile industry and focused on the number of work hours and other conditions of employment of women and children. Later an effort was made to pass legislation controlling hours of employment of adult males and then, later still, safety and health conditions.

As one might expect, the first laws were passed in the most industrialized states: Pennsylvania passed a law protecting children in 1842, New York and Illinois passed similar legislation in 1886. At the latter time, only thirty-four of the then 44 states had a minimum working age for children. By 1890, twenty-one states had statutes concerning accident prevention and provided health safeguards for employees, while eight had laws providing for factory inspectors. Massachusetts, New Jersey, Ohio, New York, Connecticut, Pennsylvania, Rhode Island, and Missouri had laws concerning workplace ventilation. By 1920, thirty-five states had laws concerned with health and sanitation provisions, some of which required "the removal of injurious dust and fumes by mechanical means" (MacClaury, 1981, pp. 18–19). A few of these labor laws included an article on the general duty of employers to protect the health and safety of employees, and these laws sometimes specified rules and codes of a more detailed character. Many states, however, did not enact any enforceable rules or laws on worker health and safety until after World War II, and most of these laws were rudimentary and poorly enforced (Fox & Nelson, 1972, pp. 45–59).

Enforcement posed a special problem. Massachusetts was the first state, in 1877, to adopt a system of factory inspections, and New Jersey and Wisconsin (1883), Ohio (1884), New York (1886), Connecticut and Maine (1887), Pennsylvania (1889), and Illinois and Michigan (1893) subsequently adopted similar systems. But factory inspectors were few in number, often untrained, and rarely professional civil servants. One factory inspector commented as late as 1940: "No factory inspection division expects factory inspectors to be chemical engineers, chemists, ventilating engineers or physicians" (Teleky, 1948, p. 71). It is only within the past twenty years that an *extensive* cadre of health and safety specialists has developed. This, as we will show, has resulted from the development of favorable legislation and has a profound impact, even though the laws themselves have subsequently been weakened.

The reality of the state system of laws, inspections, and enforcement was that—where they existed at all—they were weak, they dealt primarily with simple accident safeguards, and they featured limited inspection systems staffed by generally unqualified inspectors who rarely enforced the laws or penalized transgressors.

It is interesting to contrast the slow development of often inadequate and rudimentary safety and health laws with the comparatively rapid spread of workers' compensation statutes. By 1911 ten states had such laws (Montana was the first). By 1914 twenty-one states had compensation laws, and by 1950 all states had enacted some form of workers' compensation. Most of these laws addressed occupational accidents and did not include occupational diseases, at least until after 1945 (Ashford, 1980, pp. 47–57). The history of the development of workers' compensation is itself a fascinating story and one that requires extensive consideration on its own terms (see Chapter Six). All that can be noted here is that, in the United States, this history reflects the impact of the Progressive Era and the particular interests concerned with the potential costs of uncontrollable legal claims against employers.

Progressively greater interest in workers' compensation developed for a number of related reasons. An employee (under common law) could sue an employer for harm sustained at work, and this meant that the employer had a responsibility to furnish, at a minimum, a safe workplace or otherwise be liable for compensatory damages. In theory, these rights opened up the possibility of very large lawsuits against employers for unsafe or unhealthy working conditions. However, undertaking a lawsuit would require that employees testify against their employer—with the consequent likelihood that they would be fired. Understandably, then, legal redress was not an option undertaken lightly, especially by a largely unorganized and immigrant industrial working class. Employees would also need expensive professional legal help to stand any chance of pervailing, and few employees had sufficient resources to fund that.

Conversely, employers could readily protect themselves by claiming three common defenses: negligence by coworkers; the employee's own knowledge of the risks involved with the work; and contributory negligence by the injured worker (U.S. Congress, Office of Technology Assessment, 1985b; Eastman, 1910; Barth & Hunt, 1980; Ashford, 1980, pp. 47–49).

The disadvantages to the employee and the fact that few companies were ever forced to compensate accident victims for loss of income, let alone for physical and health damage or medical care, led Progressive Era reformers to push for a nationwide system of compensation. Employers backed their efforts, supporting workers' compensation even before organized labor did, largely because they recognized that a fixed state-run system protected them from potentially ruinous liability judgments. As C. Noble put it:

> From the perspective of business, workers' compensation had two major advantages. First, because it was intended to be a no-fault system with fixed compensation schedules, it precluded employee suit against employers. Thus it promised to cap employers' liability and, thereby, control and regularize the costs of accidents. Second, workers' compensation took conflicts over safety and health out of the workplace and channeled them into an administrative system. There workers confronted experts—doctors, lawyers and public officials—rather than employers; accordingly, the issues were redefined. (C. Noble, 1986, p. 43)

Clearly, large manufacturers were concerned that, unless these actions were taken, working-class and union unrest might force the states (or the federal government) to propose significant changes in employee prerogatives, thus contesting employers' total control of the workplace.

The major actors in the development of a compensation system were those who recognized the advantages of an institutionalized approach to the problem. The movement was led by associations that developed at the turn of the century to coordinate business strategies, deal with new labor problems, and encourage a better public image for business in response to exposes about industrial conditions and the massive profits being reaped by the oil, railroad and steel oligopolies.

The Federal Government and Health and Safety

The federal government did little to offset the weakness of this system or to promote safe working conditions. Certain trades and certain classes of employment were made the subject of limited legal measures, however. For example, a law enacted in 1833 granted compensation to disabled seamen, an 1868 law limited the workday of federal employees, and in 1908 and 1916 workers' compensation statutes were passed covering Federal, railroad, and other employees.

In only one area was the federal government a major actor, namely, in dealing with hazards in the mining industry. The Bureau of Mines, established within the Department of the Interior in 1910, was crucial to the development of mine safety legislation. However, it was not until the 1941 Federal Coal Mine Health and Safety Act that inspection authority was granted to the bureau. While the inadequacies of this legislation were clear, it was only with the passage of the 1969 Coal Mine Health and Safety Act and its amendment in 1977 that the federal government developed anything approaching complete jurisdiction and control over conditions in mining, arguably the most dangerous of all occupations.

The only other federal legislation worthy of note prior to that establishing OSHA in 1970 was the 1936 Walsh–Healey Public Contracts Act which allowed the Department of Labor to certify certain standards of safety for work done by federal government contractors. This legislation covered all employees who worked on federal government contracts. The Bureau of Labor Standards (BLS) was also given the right to inspect workplaces covered by the Act. The BLS, however, almost never used its authority: in 1969 only 5 percent of the 75,000 firms covered by Walsh–Healey were inspected, and for the 3,750 worksites visited 33,000 violations were recorded but only 34 formal complaints were issued (U.S. Congress, Office of Technology Assessment, 1985b, pp. 205–215).

Before Walsh–Healey all investigative and regulatory activities were at the stat level. These activities were largely restricted to important research being done by concerned professionals—particularly by individuals such as Alice Hamilton (founder of occupational medicine, who was subsequently appointed, in 1910, to form the Illinois State Commission on Occupational Disease). Hamilton's efforts, as well as those of Crystal Eastman (Eastman, 1910) and Morris Hilquit (head of the Amalgamated Clothing and Textile Workers Union), were supplemented by the actions of other middle-class reformers throughout the Progressive Era.

These earliest efforts at workplace reform played out against the backdrop of increasing evidence of the physical costs of unrestrained industrial development. While some major industrial firms (particularly U.S. Steel International Harvester) developed their own safety and health programs, they did so in the face of growing worker militancy—in particular, a strengthening Socialist party and an increasingly active union movement, as best personified internationally by the Industrial Workers of the World (IWW).

Further, increased concern over health and safety at work was linked to the whole public health movement (Adamic, 1963; Boyer & Morais, 1955). Fear of tuberculosis and cholera in the rapidly growing urban areas, evidence from muckraking journalists of the horrors of factory conditions, the militant efforts of, particularly, the mineworkers unions, and the attention of the AFL-CIO to such issues as tuberculosis among its members raised the occupational health and safety issue within the context of overall social reform at the turn of the century. As Rosner and Markowitz (1984) note:

> The reformers were motivated by sincere horror at the human costs of industrialization, but in addition they feared that health and safety problems would exacerbate class conflict. Reflecting the ideology of the Progressive movement, they believed that it was necessary to ameliorate certain working conditions through the cooperative efforts of industry and government; this, they believed, could go a long way toward resolving the tensions between labor and capital. (p. 467)

Many working in this movement could readily make invidious comparisons to the better conditions in Europe—especially in Great Britain and Germany, where the state had already actively intervened to regulate working conditions (MacClaury, 1981, pp. 26–31; Rosner & Markowitz, 1987; Ashford, 1980).

Towards the end of the second decade of the twentieth century these reformers found an ally in the federal government as the demands of a war economy increased the attention paid to health and safety issues, particularly those arising from production of munitions and increased use of a wide range of chemicals. Further, the U.S. government became increasingly involved in organizing production, and government agencies for directing the economy became a familiar aspect of general economic policy.

The pace and direction of reform, however, could not to be maintained in any linear fashion. As the needs of a war economy and the rationalization of production such needs provoked became dominant themes, the coalition of reformers and trade union activists became submerged in the welter of anticommunist agitation, 1920s-style political repression, and eventually the economic downturn of the 1930s. While some gains remained—the passage of extensive workers' compensation laws, for example—in general all that was left was the industry run National Safety Council (NSC). And even the NSC consistently pushed the general safety movement back into what can only be characterized as a co-opted "blame-the-worker" ideology.

The experience of the Great Depression and the militant union-ism of the Congress of Industrial Organizations (CIO) during the 1930s led to major reforms in the American welfare system, but little was done to improve safety and health in the workplace signifi-cantly. One explanation of this failure is to be found in the problems confronting the economy in general and the trade union movement in particular. That is, given the strenuous efforts to organize workers in the CIO, its subsequent twenty-year rivalry with the American Federation of Labor (AFL), and the extremely high levels of unem-ployment, it is not surprising that little attention was paid to health and safety at work. Even one of the worst cases of industrial slaugh-ter, involving the loss of hundreds of lives, the disaster at Gauley Bridge, West Virginia, in 1930–1931, did not result in the federal government's seeking new legislation (Page & O'Brien, 1973, pp. 59–60).[1] Only during World War II did the government begin to address the health and safety problem, but this effort amounted to a propa-ganda campaign that mirrored the "Safety First" movement of the 1920s and resulted neither in more stringent enforcement of exist-ing laws nor in the creation of any new federal health and safety leg-islation.

There were, however, a couple of congressional attempts to introduce legislation between 1940 and 1968, but the several bills that emerged during this period never made it out of Committee. In a number of cases, opposition to the bills focused on the encroach-ment on states' rights implied by federal legislation. Attempts to turn these bills into law fell victim to bureaucratic disputes between public health officials and labor departments at both the state and federal level (Page & O'Brien, 1973, pp. 63–64).

In 1965 the Public Health Service published a report—"Pro-

---

[1] During the building of a hydroelectric tunnel at Gauley Bridge, workers were required to drill through pure silica. None was given any respiratory protection. As a consequence, over 470 men died from the aftereffects, and 1,500 were permanently disabled, most of whom were black.

tecting the Health of Eighty Million Americans," commonly called the *Frye Report*—that formed part of the basis for the development of later legislation. This report was heavily promoted by George Taylor, a staff economist with the AFL-CIO, who became increasingly adept at using his position in the Washington bureaucracy to promote the health and safety issue (Taylor was later to play an important role in the passage of OSHAct and in the functioning of OSHA in the early 1970s).

The *Frye Report* is significant in this history because it advocated government-led social reform of workplace health and safety. Even more important was the fact that it consciously related workplace health problems to those of general public health. The report had a vision of general public health that drew the attention of health professionals to the neglected area of workplace health and safety, thus mobilizing an important segment of the scientific community around the issue. Although the report's recommendations were never adopted, Taylor's efforts drew the attention of the AFL-CIO. As a result, union leaders in Washington urged the Johnson administration to take note of the report's recommendation for an occupational health initiative centered in the Public Health Service (MacClaury, 1981, p. 21).

These developments were not, in themselves, sufficient to create a new initiative for federal occupational safety and health legislation. But they did instigate increased concern with the issue within government during this period, concern that became more manifest when evidence emerged of elevated cancer rates among uranium miners. When, in 1967, this evidence became public, it tapped a deep fear in the public about the effects of radiation exposure. The lack of action by the Federal Radiation Council prompted then Secretary of Labor William Wirtz to issues standards for radiation exposure, under the authority of Walsh–Healey. As MacClaury notes, "This move had a decisive impact on the shaping of a national job safety and health program in 1967, as the Departments of Labor and HEW [Health, Education and Welfare] promoted their competing proposals" (1981, p. 21).

Lack of action over safety and health in the 1950s and 1960s was compounded in the immediate postwar period where Cold War politics and the general success of comparative labor–management accord precluded any major attack on conditions at work. Unions themselves (with some notable exceptions) never pushed the health and safety issue until late in the 1960s, when the overall health and safety movement began to coalesce around the passage of the original enabling legislation for OSHA.

Origins of the OSHAct

The origins of the OSHAct have been analyzed extensively. In much of the literature the direct role of organized labor has been played down. Clearly, the union movement as a whole had not taken a major role in pushing for health and safety regulation. Labor, however, did react to unrest among the rank and file about working conditions and was pushed to address the issue by some individual unions during the 1960s, especially those whose members worked in the most dangerous industries.

One such union was the United Mine Workers of America (UMA), which had long been active on safety and health issues. The activities of this union were coupled with the development of the Black Lung Association and were spurred by public outrage over the devastating explosion in a West Virginia mine in 1968 that resulting in the loss of seventy-eight lives. The actions of certain members The Black Lung Association capture a sense of the forces most evident in the reform movements of the period. Their hostility toward mine owners (and often the UMA's international leadership) was expressed

> ... through both the traditional techniques of strikes and selective violence and new methods of civil disobedience. Miners and their families had both direct and indirect experience of the civic militancy of the 1960s. Military service in Vietnam and the shuttle migration to midwestern cities exposed Appalachians to the new articulateness of blacks involved in the civil rights movement. Television provided national coverage of protests, demonstrations and civil disturbances. (Fox & Stone, 1972, p. 53)

The Black Lung Association was one of three "victim's organizations" that developed during the late 1960s. The other two, the White Lung Association and the Brown Lung movement combined to form a pivotal force for raising public consciousness about the dangers to workers' health caused by airborne dust and chemicals. The revelations arising out of the work of Dr. Irving Selikoff on the impact of asbestos on worker health, the growing evidence showing that cotton dust caused serious lung problems in textile workers, and the development of a community organizations around these issues brought workers and community activists together. Local unions and unions representing these sectors began to lobby and to publicize the issue at the federal level. They also began to press their claims through the legal system and to raise the issue with their international parent organizations (Brodeur, 1973; Judkins, 1986; Berman, 1978).

The role of these victims' organizations needs further study, but the evidence suggests that they were an extremely important part of the general context of heightened awareness that developed around occupational safety and health issues in the late 1960s.

In addition to the UMA a number of other unions pushed the health and safety issue, largely as a result of the activities of individuals who took up this cause. In particular, George Taylor was joined by Anthony Mazzochi of the Oil, Chemical, and Atomic Workers (OCAW) and John Sheenan, then an aide to I. W. Abel, the head of the United Steelworkers of America (USWA).

These three men, while not acting in a coordinated strategy, were familiar with the Washington political environment and served on a number of commissions and task forces associated with public and worker health. They linked up with activists in Washington (including those associated with Ralph Nader), sought the support of the then Assistant Secretary for Labor Standards Esther Peterson, who was sympathetic (especially to the plight of OCAW members working in uranium mines, who were exposed to radon gas), and then the three lobbied their own unions to pay more attention to health and safety.

These activists began to press for a comprehensive federal law to protect workers' health. But organized labor's leadership, as a whole, was not heavily involved in the issue until after the first bills were laid before Congress; the AFL-CIO, for example, devoted little space to health and safety at its annual meetings (Donnelly, 1982, p. 14; Edwards, 1981, p. 183, cited in C. Noble, 1986). Union leaders responded only when rank-and-file activism indicated growing unrest among workers about job safety and health, or when they realized the value of finding a new issue to garner rank-and-file support.

Worker unrest, in this context, was to have a greater impact on unions than on employers or the federal government. Nevertheless, the context is important here. By the mid-1960s there were clear indications that worker militancy was increasing. In 1966–1967 the incidence of strikes was higher than at any time in the previous ten years, and the number of wildcat strikes (those not authorized by union leadership) showed a dramatic increase between 1967 and 1969 (Edwards, 1981, cited in C. Noble, 1986).

These strikes were often prompted by concern over unsafe or unhealthy conditions at work. During the period 1965–1967, for example, some 175,000 auto workers, chemical workers, and mine workers initiated wildcat strikes over safety and health issues (Donnelly, 1982, p. 17). In addition, there is evidence that working conditions were becoming a major concern among industrial workers generally. The University of Michigan's 1969 *Survey of Working Conditions* reported that health and safety hazards were considered a "sizable" or "great" problem by industrial workers (Quinn & Shepard, 1974, pp. 149–153).

These activities coincided with concern over the environment and the attempted enlistment of blue-collar support for President Lyndon Johnson's potential reelection bid in 1968; and, although the first occupational safety and health bill was not passed by Congress during the election period, it was taken up again by the newly elected Nixon administration once it was clear that organized labor had finally seized the initiative and gotten behind the bill. The story of the passage of the OSHA Act has been told elsewhere (Page & O'Brien, 1973; Berman, 1978; Kelman, 1980; Mendelhoff, 1979; Bureau of National Affairs, 1971; Mintz, 1984). It is important to note, however, that factors functioning within the government also played an important role.

As Kelman has effectively argued, the evidence seems to indicate that the initial impulse to create OSHA emanated not from the unions or concerned activists but from within the federal government itself. The Johnson administration was interested in creating a policy agenda built upon the civil rights issue. With a Democratic party majority in Congress and a new bloc of middle-class voters who were interested in social reform in general and showing increased concern about environmental and public health issues, the Johnson administration was searching for a reform initiative. The Democratic party also needed a way to cement and improve its relations with its traditional white blue-collar and union constituency (C. Noble, 1986).

As a result the administration developed a broad agenda of "quality-of-life" policies. Occupational safety and health became a portion of that agenda, in part because the Department of Labor cited it as an important issue. While some have suggested that President Johnson's chief speech writer (who had a brother in what was then the Bureau of Occupational Safety and Health) slipped references to health and safety into President Johnson's speeches, thereby drawing greater attention to the problem, it would be simplistic to accept this as the *cause* of the administration's involvement. Rather, health and safety became one of a number of policy questions developed as part of an overall concern with quality-of-life initiatives and one that could be used to boost blue-collar support for the administration.

At the level of governmental policy, then, the initiative for the law itself was generated within the administration. Pressure from trade union activists, proposals emanating from the Department of Labor and from the AFL-CIO as a result of Taylor, Mazzochi, and Abel's activities, and the inclusion of social reformers such as Nader all enabled the first bill to get a public airing (C. Noble, 1986, pp. 72–82; Kelman, 1980).

There are, however, other factors to consider in this history. Some have asserted that the OSHAct was a reaction to a rise in the trend of work accidents, but, as Donnelly notes, this perception was erroneous. Deaths from workplace accidents actually decreased from 21 per 100,000 workers in 1960 to 19 per 100,000 in 1967—a small decrease, to be sure, but obviously not an *increase*. While these figures still reflected a serious problem, they do not *explain* why the government chose to act when it did (Donnelly, 1982, pp. 18–19).

Neither was much consideration given by the federal government to increasing evidence of a major *health* crisis resulting from workplace exposure to toxic substances, despite the emerging claims of victims' organizations and the evidence presented in the *Frye Report* and in subsequent research. Others account for the pressure to pass legislation as originating in media exposure and concern, or in the growing general interest in environmental issues that some activists, implicitly and explicitly, linked to the issue of health and safety at work (Berman, 1978, p. 32; Elling, 1986, p. 395; Gersuny, 1981; Kazis & Grossman, 1982).

While these factors certainly had some impact, the general flowering of the environmental movement provides only the structural context for pressure underlying subsequent legislation. Clearly, the rank and file's discontent found fertile ground in other social and economic forces. As one author notes:

> The decision to give workers a right to safe and healthy work can be understood only in the context of the larger shift in the nature of American liberalism during the halcyon days of the Great Society. In this case rank and file discontent over work combined with the radical visions of labor activists, environmentalists and public interest reformers to feed the issue of working conditions into the swelling of demands for reform. (C. Noble, 1986, p. 69)

This shift in American liberalism was both a cause and effect of the rapid development of post-World War II American capitalism. The compromise achieved between employers and the American trade union movement and the wider impact of the Cold War and its accompanying rhetoric and ideological offensive all combined to give a degree of freedom of movement to capital. Numerous small firms developed in the 1950s to challenge the hegemony of large corporations, and capital was increasingly becoming more mobile, both within domestic sectors of the economy and internationally.

Ferguson and Rogers point out, this was precisely the time in which the New Deal coalition (not of poor whites and blacks, or millions of farmers, but the bloc of capital-intensive industries and internationally active major banks) were willing to support liberal policies in the domestic political arena. What this bloc wanted was "liberalism at home, internationalism abroad"—and this bloc effectively ruled the roost throughout the 1950s and 1960s.

While all the issues that this prevailing mindset engendered cannot be discussed here, it is clear that an increasingly powerful segment of American capital—the large multinational corporations and banks—accepted and encouraged the application of Keynesian principles to the economy. Thus, along with these policies, the powers-that-be were perfectly content to acquiesce to social programs that dealt with domestic social unrest (Ferguson & Rogers, 1986, pp. 46–55).

President Johnson's Great Society program was not at odds with the political or economic thrust of this segment of capital; nor, of course, was his administration's commitment to free trade. Where these policies ran into opposition was with the still vibrant force of domestic industrial capital and small business. The state was forced to mediate between these interests, and its efforts to do so created an opening for the class pressures slowly developing around health and safety. As we shall show, these opportunities provided for the passage of the OSHAct, but they were to be quickly reassessed when capital *as a whole* suffered enormous setbacks during the 1970s.

It is difficult, therefore, to draw a direct line from working class pressures (manifested by union activity) to the passage of federal legislation to protect workers from occupational hazards (Donnelly, 1982; Elling, 1986; Gersuny, 1981; Piven & Clowerd, 1977). The increasing militancy of the rank and file in the mid-1960s, however, did place considerable pressure on union leaders and union bureaucracy to respond in some fashion. When health and safety emerged as a dominant contemporary issue (as a result of the interaction of the various factors just described) union leadership was quick to seize the opportunities that were presented.

The history of the pre-OSHA health and safety movement would seem to indicate that "class pressures" (at least in the form of labor demands manifested in strike action or lobbying by the AFL-CIO and major unions) were not the most important factor in the passage of federal legislation. Public opinion, media exposure, and public health research played key roles in breaking open the issue. But, as Kelman notes, the existence of a set of institutions such as Congress, the Public Health Service, and the Department of Labor were very important in providing the stimulus for legislation. In this sense, the OSHAct may be said to have been generated *within the state*. Thus, state institutions were extremely important in that regard for the creation of legislation, as were the indirect effects of the whole New Deal apparatus for welfare and social policies.

The passage of the OSHAct therefore resulted from a combination of factors and the "swelling of demands for reform." Direct "class action" in the form of trade union lobbying or rank-and-file strikes and protest are only part of the story. As we have noted, the tensions within capital and the degree to which they were able to constrain and influence policy are equally important. When these factors coalesce at a given historical moment, it is then that we can see the true extent to which the state may act relatively autonomously from the influence of capitalists, or other critical social actors, around a specific policy issue (Weir, Orloff, & Skocpol, 1986, pp. 16–27).

Further, if, indeed, class-based social movements created certain pressures for regulation and combined with realignments in specific factions of capital, then can this perspective be used to shed light on what appears to be a second radical shift, namely, the advent of neo-conservatism and the deregulatory strategies of the late 1970s? Clearly, a major question is what prompts such reforms as the OSHAct. Yet, it is also important to examine the ways in which such reforms become undermined or attacked. What kinds of forces are involved here? What roles do state institutions play?

The Occupational Safety and Health Act of 1970

The provisions of the OSHAct created something of a break with past forms of social regulation. In giving wide regulatory power to the federal government through the offices of the Occupational Health and Safety Administration, it established significant legal protection for nearly every American worker. Its passage, however, was far from easy (MacLaury, 1981).

Three years of bitter legislative struggle and intensive lobbying culminated in President's Nixon signing of the act into law in December 1970. The OSHAct was hailed as a triumph for workers. AFL-CIO President George Meany called it "a long step . . . towards a safe and healthy workplace." Indeed, it did—for the first time—create a federal law mandating that each employer "furnish to each of his employees employment and a place of employment which are free from recognized hazards that are causing or are likely to cause death or serious physical harm to his employees." It also created three agencies: OSHA, to set and enforce health and safety stan-dards; the National Institute for Occupational Safety and Health (NIOSH) to research occupational hazards; and the Occupational Safety and Health Review Commission (OSHRC), to review con-tested enforcement actions.

This system, in place since 1971, has been under continual attack by both labor and industry, although labor has consistently supported a *stronger* OSHA. The passage and implementation of the OSHAct have created unprecedented political conflicts about the role of the state in the regulation of private enterprise. Clearly, however, the activities of OSHA were (and are) highly circumscribed by tradi-tional American resistance to state intervention in the economy, its vulnerability to political influence, and the overall balance of class forces in the United States since World War II.

These features also explain the genesis of OSHA in the hundred or so years since Massachusetts passed the first factory inspection law. A liberal-democratic ideology that was antagonistic to government intervention (especially at the federal level) and a weak and fragmented labor movement both contributed to the slow development of health and safety regulation and the comparative weakness of the current system.

Occupational Safety and Health in the 1970s

Even before the OSHAct was signed into law, it had become the target of attacks by business and industry. Indeed, there is evidence that the OSHAct itself was primarily President's Nixon's law—that is, a means by which he was able to seek and partially garner labor's support for his reelection in 1972. Its symbolism was important to Nixon for its electoral value.

Whatever its value to candidate Nixon, for labor the OSHA law was *vital*, representing the first organized effort to establish a federal mandate to improve conditions in American workplaces meaningfully. As we shall see, OSHA was not given sufficient resources to carry out its mandate. In fact, its solely symbolic value to President Nixon contributed to many of the ensuing problems that OSHA faced.

Conflict and compromise marked OSHA's performance during its first few years of operation. Little effective action was taken by OSHA to deal with the health and safety problems of American workers, with major conflicts arising from the very beginning. By immediately adopting the comparatively lax 450 health standards used by the old Bureau of Labor Standards, which had been largely developed by private industry over several decades (based on little or no scientific evidence), the agency effectively positioned itself to be *unable* to control health hazards in American industry.

Enforcement was another area of weakness. Few inspectors were recruited to the new agency, and the fines imposed during this period had little deterrent effect. The average fine was less than $50, and the maximum for serious violations of OSHA standards averaged approximately $625.

More importantly, OSHA's record in setting and enforcing new standards was very uneven during this initial period. OSHA adopted only three new health standards in its first three years, plus the never used package regulation on fourteen suspected carcinogens. By 1976, OSHA had issued four new standards, but mainly in response to lawsuits lodged by unions. From 1971 until 1984, OSHA issued a total of eighteen separate health standards and twenty-six safety standards.

In these first years OSHA did little to deal with health and safety violations on a large scale. In fact, there is clear evidence that OSHA targeted its activities primarily at the small-business community. This makes in that some small firms may tolerate truly dangerous conditions, particularly if they have neither sufficient resources nor a proper concern for public opinion and public welfare to act on their own to improve health and safety conditions. Yet, it is also true that small firms do not in general wield sufficient political clout to lobby against such regulation and are therefore somewhat "easy" targets.

This initial small-business focus ended up having negative political consequences. By focusing on small business, OSHA antagonized a group that proved pivotal in a subsequent squaring off against OSHA. Small businessmen were the most vocal proponents of entrepreneurial freedom from government regulation. As such, they provided a crucial ideological component in the development of antistate, antiregulatory populism.

Despite the small general improvements in health and safety in the 1970s, the 1974 oil crisis and ensuing rapid economic downturn put increasing pressure on American business and industry and was to have significant consequences for the ability of the developing regulatory apparatus to continue effectively.

The worldwide recession that followed the first oil crisis had a major economic impact in the United States. The unemployment rate rose rapidly, from 4.9 percent in 1973 to 8.5 percent in 1975, and the gross national product fell in both 1974 and 1975. By 1975 the inflation rate had reached 9.1 percent. As the economic situation worsened throughout the mid-1970s, the political climate became less open to the kinds of social regulation popular at the beginning of the decade. As a result, business and industry had to find a way to overcome the setbacks they had suffered at the hands of regulation (Szasz, 1984, pp. 108–109). To do so, they reorganized their lobbying efforts and attempted to equate the profitability of individual firms with the continued health of the overall economy—in particular, the maintenance of standards of living.

This reorganization was to have major implications not only for OSHA but also for the overall relationship that business was to have with the federal government. As Noble points out, the shift was subtle:

> Many of the most important firms and interindustry groups ceased to deny the right of the state to regulate markets or the reality of the health and safety crisis. Instead, they defended their interests in more subtle ways. Most important, employers attempted to identify their particular interests in lower costs and higher profits with a general societal interest in jobs, economic growth, and capital investment. Economic growth, business suggested, was not only important as protection but was a precondition for it. (C. Noble, 1986, p. 104)

Despite the best efforts of the business lobby, however, the late 1970s bore witness to one of the more upbeat periods for OSHA, that is, during the Carter administration. Many of the regulatory agencies benefited from the activism of Carter appointees, and this was especially true of Eula Bingham, an academic and professional industrial hygienist who became the director of OSHA. Under her brief tutelage, OSHA sought reductions in permissible exposure levels to toxics and resisted efforts to apply cost–benefit analysis to standards. In 1978 OSHA issued six major new health standards and challenged economic review of those standards. Bingham also pushed the agency to increase and rationalize its enforcement strategies; trivial standards were removed, fines were increased, and follow-up inspections of plants were increased. OSHA during this period attempted to increase worker health and safety rights and to develop worker-oriented training programs (under the New Directions Program). This brief period of a proactive OSHA between 1978 and 1979 ended as the Carter administration increasingly gave in to increasing pressures from the business lobby, particularly following the onset of the Iranian hostage crisis in November 1979.

Occupational Safety and Health in the 1980s

The 1980s embodied a more pronounced drift to the right, exacerbating the rightward tilt already evident during President Carter's final two years in office. The neoconservatism of the Reagan–Bush years has had a profound effect on a variety of contemporary social and political circumstances. How, then, did OSHA and occupational safety and health regulation in general fare? The effort to control federal regulation of working conditions was perhaps the most successful thrust of the neoconservative movement in the United States. By using the set of White House oversight controls already established in the 1970s, the Reagan administration was able to move rapidly to, effectively, "gut" OSHA.

The administration's first action was to impose a sixty-day freeze on all new regulations. Almost immediately the White House set up a regulatory task force under Vice-President George Bush's chairmanship. However, the central thrust of this policy was Executive Order 12291, issued in February 1981, which gave the Office of Management and Budget (OMB) the power to oversee all major regulations. If new regulations were issued, all regulatory agencies were required to present extensive cost–benefit analysis to justify them. OMB became, therefore, a major actor in the struggle to decrease social regulation. President Reagan quickly made OMB *the* regulatory manager. The combination of OMB's oversight, a variety of regulatory review bodies, and presidential executive orders gave the White House enormous control over many of the regulatory agencies, particularly OSHA.

A second strategy to cement this control was to place individuals friendly to the administration's political agenda in positions of control in regulatory agencies. While President Reagan was unable to achieve his campaign promise to do away with OSHA, the combination of political appointees and the regulatory review efforts were sufficient to remove the last vestiges of power from an agency already weakened by the attacks on it during the 1970s.

These strategies controlled social regulation in three ways: (1) economists became important to regulatory policy making, and critics of regulation were appointed to economic policy positions; (2) it became increasingly difficult for nonbusiness interests to get their voices heard because the review groups and the OMB were not easily accessible to public input; and (3) the requirement for cost–benefit analysis of regulations changed the nature of how regulations were to be assessed—that is, they were now to be evaluated primarily in terms of macroeconomic effect and not in terms of their impact on the health and safety of workers (C. Noble, 1986, pp. 159–160).

SOCIAL REGULATION IN THE 1980s:
THE COLLAPSE OF OSHA

The business agenda that was shaped in the 1970s came to fruition at the start of the new decade. Beginning with the 1974 recession and the first oil crisis, business recognized that it had to create a coordinated strategy for restoring long-term profitability. The reorganization of business lobbying groups, initiated within one or two years of the passage of OSHA, has been described and analyzed more fully elsewhere (C. Noble, 1986). In the case of OSHA, the main thrust of business lobbying was directed at the White House.

As a result, a succession of presidential controls effectively throttled the regulatory impulse at OSHA. President Ford issued the first relevant executive order, requiring regulatory agencies to submit Inflation Impact Statements to the OMB. He followed up by creating the Domestic Council Review Group on Regulatory Reform. Subsequently, the Carter administration required Economic Impact Statements [EIS] for all new regulations (including but not limited to OSHA and further cemented control of regulatory agencies at the White House by forming two new agencies: the Regulatory Analysis Review Group (RARG) and the Regulatory Council. As these agencies were organized through the Executive Office of the President (EOP), they provided intense scrutiny of regulatory bodies—particularly those concerned with occupational health and safety (OSHA), the environment (the Environmental Protection Agency) and consumer protection (the Consumer Product Safety Commission).

This scrutiny was aided considerably by the role of budgetary oversight. Beginning with Gerald Ford's Executive Order 11821 (issued in 1974), which gave OMB the right to review all regulations that might be construed having an inflationary impact, the White House asserted its control over much social regulation. The role of OMB, as will be discussed later, became increasingly important. In combination with the Council on Wage and Price Stability, the Regulatory Analysis Review Group and the Regulatory Council, the Office of Management and Budget (and with it the Executive Office of the President) gained greater and greater control of the regulatory agencies under its purview.

White House oversight and the need to meet economic criteria for its standard-setting procedure meant that, not only were the mechanisms for deregulatory policies in place, but also OSHA faced considerable constraints on its decision-making process. While the agency tried to resist these constraints during the latter part of the 1970s, the constant threat posed by regulatory oversight significantly decreased its autonomy and made pursuit of its proper goals exceedingly difficult.

The effects here are twofold. The agency's role is reduced to a minimum while at the same time the agency loses credibility with both employers and employees. The agency no longer has the authority to command compliance, nor is it trusted by workers themselves. The result is that OSHA itself becomes merely symbolic. This symbolism reflects the structuring of the agency's role from its very inception. Like much social legislation in the United States, the OSHAct has satisfied the ideological requirement that the state *appear* to be acting in the general interest.

These effects could not have been achieved without the groundwork that was already laid during the Ford and Carter administrations. This was especially true of the oversight rights that gradually accrued to the White House. Placing OSHA directly under OMB purview made it much easier for successive administrations to weaken the agency and therefore the content and import of the OSHAct itself. This process was further facilitated by the ability of the president to nominate a personal candidate to head OSHA. This appointment process proved important to the Reagan administration's efforts to curtail regulatory agencies and was highly effective in a number of areas of social regulation.

OSHA IN THE 1990s

The 1990s began with a profoundly weakened regulatory system for occupational safety and health. While in the late 1980s OSHA did increase its activity in a small number of high-profile cases (a Phillips Petroleum plant that exploded in 1989 resulted in a $5.7 million fine; USX [U.S. Steel] was hit with a then record $7.3 million fine in the same year), these were still very much the exceptions. In 1990 Congress authorized a sevenfold increase in maximum fines and established a $5,000 minimum for willful violations, but these "upper-limit" fines are still rarely invoked, in practice. Indeed, OSHA has consciously developed a policy of avoiding punitive fines against employers (C. Noble, 1992, p. 46).

Although our discussion of the politics of OSHA over the past twenty years or so may seem abstract; the failure to regulate has very real impact on human needs and dignity. Take the case recently reported by the Associated Press:

If you have to go, the government says the boss must let you. The federal agency that oversees workplace health said Thursday that employers not only have to provide restrooms, they have to allow workers to use them.

For some workers, that's not always been the case.

The Labor Department's Occupational Safety and Health Administration has long required that toilet facilities "be provided in all places of employment" for all workers. But that regulation merely requires employers to have enough bathrooms. It says nothing about giving workers access to them.

There's no problem for most of the nation's workers; they just get up and go. But in some jobs, such as food processing, assembly lines, and telemarketing, meeting a simple human need can involve pleading and even the risk of losing a job.

OSHA spokesman Stephen Gaskill said the agency believed that when it required restroom facilities in all workplaces 20 years ago, it had made clear to companies that workers had the right to use them as the need arose.

Now, he said, there have been enough complaints to show that the rule must be spelled out.

"We're telling employers that within reason and when necessary workers should have the ability to use a restroom," Gaskill said.

In a memo interpreting the previous regulations, the agency said it is clear that "all employees must have prompt access to toilet facilities."

"Restrictions on access must be reasonable and may not cause extended delays," the memo said. "Timely access is the goal of the standard."

It noted that a number of companies concerned about uninterrupted production have set up signal or relief systems for workers on assembly lines or other jobs "where an employee's absence, even for the brief time it takes to go to the bathroom, would be disruptive."

"Under these systems, an employee who needs to use the bathroom gives some sort of a signal so that another employee may provide relief while the first employee is away from the work station," the OSHA memo said.

"As long as there are sufficient relief workers to ensure that employees need not wait an unreasonably long time to use the bathroom, OSHA believes that these systems comply with the standard," the agency said.

The OSHA memo stated that workers denied prompt access to restroom facilities can suffer adverse health effects. OSHA field inspectors can issue citations for violations. (Knutson, 1997)

The Clinton administration seemed to offer a return to more liberal politics after over a decade of conservative ideology, but little was done to stem the decline in OSHA during its early years. Funding for the agency and (its research arm) continued to be cut, and, despite claims by then Labor Secretary Robert Reich that President Clinton was committed to worker safety, the health care reform proposals of 1993–1994, the constant diversion of attention to scandals, and then the election in 1994 of a massive Republican majority in Congress all but killed the agency. Despite the introduction of several bills to reform OSHA in 1993 and 1994 and despite support from several major unions, the AFL-CIO, and a number of environmental groups, the coalition to reform and strengthen OSHA was simply too weak. The reform proposals have languished, and there is little likelihood that they will succeed in the near future.

A similar fate befell the major proposal on an ergonomics standard proposed by OSHA in 1995. The proposal would have affected control and costs in nearly all private workplaces in America. As a consequence, it was met head-on by a powerful coalition of business interests (led by the United Parcel Service) that defeated the proposed standard.

THE POLITICAL IMPLICATIONS

The creation of numerous forms of social regulation in the late 1960s and early 1970s has been seen by many commentators as a setback for capital (Kazis & Grossman, 1982; Green & Weitzman, 1981). Such regulation not only necessitated large capital outlays to meet demands for engineering controls for pollution and worker safety but also challenged capitalist control at the point of production. Clearly, a whole range of legislation concerned with protecting workers, consumers, and the environment (created in the early

1970s) could be regarded as a major attack on the prerogatives of industry.

The politics of neoconservatism, however, grew out of the crisis of world capitalism in the 1970s and not simply as a reaction to the pressures of organized labor or the environmental and consumer movements. The end of consistent economic growth that had marked the immediate postwar era ushered in a period of unprecedented social and political unrest—exacerbated, in part, by the Vietnam war—and created new coalitions of actors and increasing concern with citizen rights and environmental issues.

As these forces developed momentum, the two major recessions of the 1970s simultaneously undermined the essential class compromise that represented general acceptance of welfarism and provided fertile soil for advocates for the deregulation of regulated private industry. In Europe, and particularly in Britain, the deepening economic crisis portended the breakup of this compromise. The loss of faith in the welfare state and in Keynesian principles of economic management to effectively control and sustain economic growth became increasingly widespread. These doubts, and the erosion of confidence they engendered in the state, brought a marked change in political attitudes by the mid-1970s. Business and industry was not slow to pick up on the import of these developments, especially as a real crisis in capitalism took hold in Europe and the United States.

In the United States the overall weakening of America's world dominance exacerbated these changes. The relative collapse of American hegemony in global politics and, particularly, the loss of foreign and domestic markets to European and Japanese capital, as well as the internal threat posed by what appeared as an aggressively regulatory American state, combined to raise serious problems for capital as a whole. Faced with these setbacks, capitalists organized a response that went beyond piecemeal attacks on specific policies and orchestrated a change in the *politics* of doing business in the United States. This, we would assert, provided the direct impetus for creating a conservative political environment, one consequence of which was the eventual election of the Reagan and Bush administrations.

On three fronts, the politics of the Reagan–Bush era have proved extensively beneficial to capital. First, government tax and monetary policies resulted in a massive redistribution of income toward the rich and to corporate America. Second, as we have noted, there has been a concerted attack on union power that has had the effect of reducing workers' bargaining positions and has delegitimized unions as political and economic actors. Third, the promotion of antiunion, antistate, and anticommunist ideology has given a mask of legitimacy to the neoconservative movement, thus facilitating its more widespread acceptance.

Further—and most important for the argument being developed here—the declared policies focused on freeing American industry from government regulation. This latter thrust of neoconservatism was enormously aided by the groundwork laid by previous administrations and was facilitated by the political climate developed in the ideological promotion of laissez-faire attitudes.

This attack has taken a number of forms, but its antecedents created firm foundations for the success of the movement. It was facilitated by the lack of opposition to the Reagan administration's policies. The Democratic party was in disarray after what was, in the American system, a massive 1980 presidential election defeat for the party, subsequently repeated in 1984 and 1988 (Ferguson & Rogers, 1986). In addition, the regulatory agencies had been weakened by public opposition engendered by charges of corruption and mismanagement and fed by business attacks on their activities throughout the mid-1970s. They were thus vulnerable to the campaign that the Reagan–Bush administrations mounted against them.

These policies were aided by anti-Soviet and anticommunist propaganda that during the Reagan period (while not particularly new to American political culture) took an especially virulent form. Reagan's repeated characterization of the Soviet Union as the "Evil Empire," attacks by the religious and extreme right on socialists, communists, and gays, and, of course, the ongoing offensive against the trade union movement—all of these populist appeals were wrapped in American nationalism and packaged in a form that strongly recalled the worst of the McCarthy period.

CONCLUSION

The decline in government regulation and enforcement of workplace health and safety must be judged in the light of this recited history. Clearly such forms of social regulation as state control of workplace hazards, while perhaps contributing to the "weakness" of capital, did not cause it. Similarly, the emergence of neoconservatism pre-dates the Reagan administration. The policies of that administration cannot be judged, therefore, within a purely instrumentalist frame-work (i.e., that they are solely the acts of individuals responding to immediate issues). Rather, we would argue, the context and policies described above are a response to a variety of political and economic crises that came to a head in the 1970s. The result has been a politi-cal strategy that deliberately restricts the power of the state in order to redress the balance of class forces in favor of capital. Domestic and global changes in capitalism's development suggest that, by the 1970s, the state was no longer able to *integrate* the demands of work-ers with the needs of capitalism.

Within the context just discussed, organized labor in the United States has been forced to relinquish the initiative it developed around health and safety in the 1970s. This development has meant that labor is no longer positioned to protect workers on the shop floor or in the office.

The coming to power of the Reagan administration was the out-come of significant changes in the political economy of the United States since World War II. In periods of economic expansion such as the 1950s and 1960s when the United States, as a world capitalist power, was able to accommodate to domestic pressures for reform, capital as a whole had little need for expressly right-wing domestic policies. At the same time, the incorporation of the labor movement into American business allowed for increasing profitability even as it allowed for major labor and social reform.

The progressive movements of the 1960s were able to use the relatively unique circumstances of that period to press for ameliora-tive social legislation. But, by the end of the 1960s, the costs of the Vietnam Warm the competitive challenge of Western Europe and Japan to markets once dominated by U.S. firms, and the quadrupling of oil prices in 1973 and 1978, threatened U.S. commercial domi-nance. In particular, the collapse of the international monetary regime (signified by Nixon's devaluation of the dollar in 1971) forced American capital to unify politically and ideologically. Throughout the 1970s it did just that.

By regrouping politically, American capital was able to take back the initiative partially lost during the preceding decade. As the polit-ical and economic crisis deepened, capital was also able to attack the power of the labor movement and exploit its decline. In this it was enormously aided by the structural changes rapidly affecting the industrial landscape.

One consequence of this complex of developments was to enable capital to mount a sustained critique of government regulation of the economy, not the least of its targets being legislation to protect workers from occupational health and safety hazards. The state responded to these pressures and organized the discourse and policies of deregulation, building on structures created by previous administrations. The legacy of these developments is the rightward drift in American politics, exemplified in administrations from that of Reagan to that of Clinton, because it has entailed the full embrace of neoliberal policies in the United States and elsewhere even under the slightly more progressive social agenda of the Clinton administration—resulting in greater and greater social and economic inequality. The pernicious effect on the health of workers has been dramatic.

CHAPTER SIX

The Politics
of Workers' Compensation

A major issue attending any discussion of the political economy of the work environment is the way in which workers get compensated for illness and injury caused by their employment. The workers' compensation system in the United States has some important and unique features. Understanding its origins, functions, and implications for worker health and safety is critical.

Between 1972 and 1992, the yearly cost of workers' compensation programs for injured workers rose from $6 billion to $60 billion, representing a compounded annual increase of 12.5 percent (Burton, 1994). This is probably an underestimate, since there are many large companies that "self-insure," setting aside funds to cover any liability incurred but not paying insurance premiums. Workers' compensation is big business—it is finance capital at its finest. The injury to the worker that occurs at the point of production is commodified and traded, it becomes a thing of the market and the source of fortunes for the owners of the insurance companies, but, even more importantly, the source of great financial power for the managers of the insurance carriers. On the one hand, workers' compensation programs are literally about the pain and suffering of workers; on the other, they are one more way in which capital accumulates and capitalists rule.

The following press release from A. M. Best Co., a company that reviews and rates insurance company financial performance, gives one a view into the financial realities of a major worker compensation carrier.

WORKER INJURY BIZ BOOMING

OLDWICK, N.J.—April 8, 1998—Effective immediately, A. M. Best Co. has upgraded the "A" (Excellent) rating of Liberty Mutual Insurance Companies, Boston, to "A+" (Superior). The rating applies to the group's 11-member pool, led by Liberty Mutual Insurance Co.

The upgrade reflects the group's outstanding capitalization, conservative balance sheet, successful risk-mitigation and business diversification strategies, excellent operating performance and dominant market position as the nation's largest workers' compensation insurer. Liberty Mutual's franchise is further enhanced by its well-regarded service reputation, strong client relationships, high business retentions and effective, low-cost distribution network. In addition, the group's extensive unbundled service capabilities, risk management services and strategic alliances with managed-care networks provide for significant competitive advantages and its superior market profile.

Liberty Mutual has considerable financial flexibility through its access to capital markets. Since 1995, the company has raised $1.2 billion in additional capital, including a 100-year $500 million surplus note in 1997. Liberty's capital strength is bolstered by its conservative reserving philosophy, significant economic values accumulated in its core reserves and favorable asbestos and environmental funding and claims-mitigation strategies. Finally, Liberty benefits from its proactive management and mutual ownership which enables the group to pursue long-term growth strategies without being constrained by short-term earnings pressures experienced by many of its stock peers. (*Business Wire*, 1998)

THE WORKERS' COMPENSATION SYSTEM

In the nineteenth century, if a worker was seriously injured on the job, the only way he or she could recover any financial compensation from the employer—indemnity for lost wages, medical care costs, or any other consequences of disability—was to sue the employer. Most workers did not sue—most were simply pauperized. However, if a worker did sue, employers had three defenses in the common law—defenses that were successful in depriving workers of compensation: (1) worker assumption of risk, (2) contributory negligence, and (3) the "fellow servant" rule (Barth & Hunt, 1980; Lubove, 1967).

These arguments provided an effective defense for employers—and in some ways partook of the rationality of preindustrial production. After all, the highly skilled artisans of yore did know at least as much as the employer about the risks of their jobs—the highly skilled journeyman had much control over the labor process, and the courts respected that control. With the rise of mass production, however, workers lost control over the labor process—in fact, that was much of the point of industrialization and, later, of Taylorism and Fordism.[1]

Between 1870 and 1920 the massive scale and modern organization of production characteristic of what we today call capitalism was put into place in the United States. And with that shift in control and that change in scale, workers became subject to an industrial despotism that belied the common law notion of the hardy yeoman/journeyman/artisan. Perhaps even more important, the rising deskilled but enfranchised proletariat gained some political power and succeeded in getting "employer liability" laws passed. These laws were simple: they simply renounced the three common law defenses of the employers. The results were swift: those workers who went to court began to win their suits against employers (Croyle, 1978).[2]

From this new situation emerged what is now called an historic compromise. Beginning in 1911, state legislatures began to pass workers' compensation laws—no-fault laws that provided wage replacement and medical care expeditiously for injured workers and avoided extended legal battles. The laws also barred employees from suing their employers. The indemnity payments provided some portion of the wage but did not include any recovery for pain or suffering. The system also kept these issues out of the hands of juries, substantially reducing the uncertainty about the cost of accidents that employers faced when confronting twelve ordinary citizens. Of

---

[1] See Chapters Three and Four for a more detailed discussion of Taylorism and deskilling.

[2] There is some evidence that just before passage of workers' compensation laws some employees were successfully suing their employers.

course, many workers did benefit from these new laws, in that they either would never have sued their employers or they might very well have lost in court.[3]

An essential part of the new workers' compensation laws was that they mandated that employers insure themselves against the risk of industrial injuries. In some states, exclusive state insurance funds were established; in a few, competitive state funds were set up; but, in most, employers were told to purchase insurance from private carriers. This permitted a pooling of risk among employers, with each employer paying premiums that could be treated as any other cost of production. In this way, the vagaries of the tort system were eliminated.

Nevertheless, one aspect of the appeal of the new workers' compensation system was that it would provide market incentives for industrial safety. Premiums were to be based on the accident experience of employers, thus penalizing unsafe employers and industries while rewarding safe ones. Note, however, that these were to be relatively minor adjustments compared with the dramatic uncertainty built into the awards of the jury system. In any event, to this day there is substantial debate over the effectiveness of injury prevention through workers' compensation. If it does serve a precautionary role today, the casual effects must be slight because of the difficulty in teasing them out (Boden, 1995).

By the mid-1920s almost every one of the states had adopted workers' compensation statute. This state-imposed safety net was to be the most basic element in the system for worker health and safety in the country for the next fifty years.

The New Role of the State

Workers' compensation was the first of the social insurance schemes borrowed by the United States from Europe. Indeed, the first workers' compensation system was created in Prince Otto von Bismarck's Germany in 1884, and Britain and France quickly followed suit, developing systems of compensation in the 1890s and early 1900s (Dwyer, 1991).

The compensation scheme was a way to use the state to assuage working-class discontent and yet preserve the integrity of capitalist

---

[3]The origins of the workers' compensation system are also discussed in Chapter Five.

control of the production process. It also removed industrial injuries from public view, institutionalizing the conflict between capitalists and workers. The state, by embracing this system, mediates this conflict and reduces it to a legal ritual. At a more general level, as Dwyer notes:

> . . . it would reduce what appeared to be a growing area of worker dissatisfaction and the labor movement's organization and power and it would do so without impinging on the organization of employer power within the workplace. This would be done by building up a system of standardized payouts for technically standardized injuries. . . . In making such a definition, the institution would define itself as "neutral." (Dywer, 1991, p. 32)

Certainly firms should be encouraged to reduce accidents, but the unavoidable toll could be treated as an "externality" and a social mechanism put in place to ensure social stability. The solution that emerged was certainly not the only way one might deal with the hazards of mass production technology. In Massachusetts, in the early part of the twentieth century, for example, the state legislature was persuaded to ban the "kiss of death" shuttle (it was called this because it required workers to use their mouths to thread the shuttle) from the textile industry because of fears of tuberculosis and a potent political alliance of labor, public health officials, and the Catholic Church. That is, the state decided to intervene in the choice of technology, limiting capital's freedom to manage its own property. The application of such direct state intervention more broadly would have put state governments at loggerheads with business interests. Workers' compensation represented a less fractious alternative solution to the problem.

The coalition that was built in Massachusetts represented a counterhegemonic historical bloc: labor, public health, and church united to use the state to limit capital's control of technology. The overriding idea was that the state could and should intervene to prevent occupational disease. But this intervention occurred only because of the power of this bloc to make demands on the state.

Workers' compensation represented a different solution to occupational injuries for a primarily capitalist hegemonic bloc that included finance capital and insurance companies, manufacturers and railroad owners, and some segments of the labor movement and social welfare movements. The immediate rewards justifying this solution were expedited income replacement and medical care for injured workers, with prevention of subsequent injuries occurring through natural market mechanisms. It was a compelling idea that found ready acceptance in the United States, as it had in Europe, as an alternative to the financial and political costs of an unregulated, uninstitutionalized system of worker rights.

When a Worker Is Injured

The workers' compensation system is not only an important financial phenomenon that is a means for rationalizing the capitalist state; rather, it is also a system in which real workers who are really injured or sick as a result of their employment try to get income for the survival of their families and to pay their doctors (Schwartz, 1993). Today, in the United States, this "no-fault" system is usually the only place an injured worker can go for some kind of economic help.

Here briefly is how the no-fault workers' compensation system operates in Massachusetts, a state where only private insurance carriers may undertake that function. When a worker is injured on the job, the employer must file a report of the injury to the insurance company and to the state workers' compensation oversight agency, the Division of Industrial Accidents (DIA). The insurer must either accept or reject the claim within fourteen days. Workers' compensation requires that carriers pay for all medical expenses resulting from workplace injuries. Doctors are paid directly by the insurer according to rates set by the Massachusetts Rate Setting Commission. Hospitals or doctors may not charge workers additional amounts.

If a worker is temporarily totally disabled, the weekly benefit rate is 60 percent of the workers average weekly wage (up to a maximum of the average weekly wage for the state at the time of the injury). But qualifying for cash benefits requires that the injury prevent one from working for six days or longer. Only when the injury causes more than twenty-one days' loss of work will a disabled worker be compensated for the first six days. In Massachusetts, in 1992–1993, the maximum rate that an injured worker could receive was $543.30 per week. Fringe benefits generally are not included in the calculation, nor are they compensated for. On the other hand, workers' compensation benefits are also not subject to income taxes.

In Massachusetts, temporary total disability benefits may be paid for up to three years; after that, if still unable to work, the worker must ask to be designated permanently disabled.

If a worker is injured on the job but is still able to do some work, he or she may be designated "partially disabled." In which case the insurer pays 60 percent of the lost earnings up to a maximum of 75 percent of the total disability rate. For example, Robert Schwartz describes a case in which the worker

> earned $600 per week in a factory job. He suffered a serious back injury. He was treated for one year and received total disability benefits of $360 per week. He then found a store job doing light work. The store job paid $300 per week. Because of his wage loss, he is entitled to continued compensation benefits from the factory insurer equal to $180 per week. (Schwartz, 1993, p. 19)

The "partially disabled" benefits are available for a maximum of five years, with a few exceptions. For example, a worker who has lost 75 percent of the function of an eye, hand, leg, or foot, or who has a "permanent life-threatening physical condition," or who is permanently disabled by occupational disease, can receive the partial benefits described for up to seven years (Schwartz, 1993, p. 20).

Workers who are permanently and totally disabled—that is, those who not only cannot do their old jobs, but cannot do even light or sedentary work, who never expect to work again—are eligible for two-thirds of their pay (subject to the maximum of the state weekly average) for the rest of their lives. There are also annual cost-of-living adjustments, but frequently these are not paid if the worker is getting social security disability payments or similar benefits. Disabled workers are also permitted to earn nominal amounts from flea markets and the like.

Additional "benefits" of the workers' compensation system include special payments for "loss of function" and scarring. Schwartz writes that "each part of the body is rated. The rates are multiples of the average weekly wage in the state at the time of injury. On October 10, 1992, for example, the benefit for total loss of lumbar back function was $17,385." One-line disfigurement benefits are calculated from guidelines providing, for example, (at the high end), "FACE Wide scar, with discoloration . . . 6.5 × SAWW (State Average Weekly Wage) per inch" to (at the low end) "HAND Linear

Scar, no discoloration . . . 1 × SAWW per inch" (Schwartz, 1993, p. 21).

If a worker is killed in a job-related accident, the surviving spouse is eligible for a benefit of two-thirds of the decedent's wages, up to the maximum in effect, with a minimum of $110 per week. These benefits terminate on remarriage or after five years if the survivor has found employment. Any minor children are guaranteed $60 per week if the parent becomes ineligible for benefits. A maximum of $4,000 is provided for burial of the worker.

The Massachusetts DIA will also provide vocational counseling and may require an insurer to pay for worker retraining, including college tuition and books.

Finally, in egregious cases the DIA may award an injured worker double compensation. Such awards are infrequent since the worker must show that the employer deliberately violated safety standards or knowingly caused the injury. Double compensation is normally awarded whenever minors are injured and whenever state child labor laws have been violated.

The Importance of Class

In New York State recently, as in many other states, business and insurance carriers have sought to use a purported crisis in workers' compensation costs to force a rollback of benefits (Tarpinian, Tuminaro, & Shufro, 1977). When advocates noted that there was evidence that the insurance carriers were overcharging companies relative to the cost of the benefits, the New York State business council responded that this was just a matter of cost shifting, not cost reduction. Implicit in this response is a finely developed class consciousness: employers and insurers are of the same class and know it; they may have their squabbles, but that is really an internal matter. Cost reduction means taking it out of the hide of the working class, whether that class is conscious of itself or not. Thus, although workers' compensation in traditional liberal ideology represents an "historic compromise," the struggle between labor and capital continues in an arena dominated by capital.

In the Massachusetts workers' compensation system, the provision of medical benefits, the partial compensation for wage loss, the rehabilitation benefits, the payments for scarring and disfigurement and loss of function, and the survivor's benefits—all of these are conditioned on the employer's actually filing the first report of injury and the insurer's accepting the case. Note the attendant potential employer sanctions: if an employer fails to file first reports three times in one year and the state discovers it, the resulting fine is $100. The insurance carrier is required to investigate and make its findings available to the injured worker within fourteen days after receiving the first report. If it fails to communicate with the worker within this time period, with an explanation of its decision, it must pay the worker an additional $200.

Naturally, both the insurance companies and frequently politicians are quick to cite potential fraud as one cause of escalating costs. The media (particularly the broadcast media) are ever ready to seek out workers on workers' compensation who nevertheless are able to make the rounds of a golf course or even make their way to a dance floor. The images that are projected are both shocking and deeply ideological. Thus, media reports typically overstate the incidence of malingerers that cheat on the system while directing little attention to genuinely injured workers and the great difficulty most workers face in getting even the minimum compensation that the system affords.

In the public mind, fraud in the workers' compensation system is automatically equated with fraud committed by the worker—the unscrupulous little guy who suffers questionable harm and is chiseling or "working" a system that lacks sufficient checks and controls. Only rarely do the media focus on fraud by employers, who have much to gain on their bottom line by reducing or avoiding the cost of their harm to workers and their families. The employee is targeted as the visible beneficiary of the compensation system; he or she has standing in the system as the party who has something to gain. The complexity of workers' compensation operations serves to camouflage the fact that employers gain, sometimes substantially, when they can reduce their obligation to pay out. Therein lies an incentive for institutional fraud. Because that motivation is not readily apparent, however, employers involved in campaigns to "reform" workers' compensation can divert attention from their roles in the system and refocus the public eye on the shortcomings of the system and its other actors. Such "reform" campaigns often end up reducing injured workers' benefits as a way to curb runaway costs. And the public generally remains a passive bystander because it fears loss of local jobs, knows little about the system, and does not perceive its own stake in it.

If the media have virtually ignored potential fraud by employers, social scientists, too, have avoided the issue. Very little research—quantitative or qualitative—has examined employer fraud in the system. The question of widespread systemic fraud by employers is a rarely considered possibility, even though employers face tremendous pressures to control costs and constantly increase company profits. Even worse, if employer fraud is widespread, it may seriously distort and undermine the whole workers' compensation system. It also may displace the costs and burden of workplace injury, illness, and death onto other social services systems.

What do we mean by "fraud"? Employer fraud may take any of several forms, but all will seek to lower costs to the company, often by avoiding insurance altogether or by seeking to reduce premiums. More specifically, employers may:

- Ignore their obligation to obtain workers' compensation insurance coverage;
- Misreport the size of their payroll;
- Seek to circumvent the employer–employee relationship that generates obligations—either by leasing employees or moving to outside contractors (particularly in the construction industry);
- Misstate an employee's job classification (inasmuch as different jobs rate different insurance premiums);
- Draft a false job description after the occurrence of an injury;
- Illegally obstruct payment of legitimate claims;
- Fail to report injuries and try to "buy" employees' silence through threat of job loss or with outright cash payments.

A number of states moved legislatively in the early 1990s to curb spiraling costs in the system; many of them chose to institute measures against fraud:

- Some established fraud units within their workers' compensation systems, insurance departments, or offices of state attorneys general (states taking such action in 1993 included Arkansas, Louisiana, Kansas, Montana, and New Hampshire);
- A number of states changed workers' compensation fraud from a misdemeanor to a felony offense (Arkansas did so in 1993, Virginia in 1994) or from a lesser misdemeanor to a greater one;
- Some states increased penalties (Hawaii and Virginia, 1993);

- A few states criminalized penalties (as happened in Louisiana and Oklahoma in 1993);
- Some states set procedures or penalties for uninsured employers (for example, Georgia in 1994) and underinsured employers (Montana, 1993);
- Some states tightened the definition of the employer–employee relationship or tackled conditions for independent contractors or leasing companies (Colorado, Nevada, Oklahoma, Oregon, Rhode Island, and Utah). (Burton, 1994)

Such controls are relatively new. It may be premature to expect them to yield valuable data instantly about abuse within the system. But the data collected by fraud units should soon be available to provide a more detailed picture of the extent and forms of employer fraud. For instance, since 1993, a Rhode Island Labor Department investigator has caught 136 companies that carried no workers' compensation insurance. Although the firms were cited, state policy has been to skip penalties against them so long as the offenders agree to obtain coverage. The Connecticut fraud unit investigated 834 charges against companies that failed to provide coverage during fiscal years 1992–1994. Despite these numbers, the state collected only $15,000 in civil penalties for noncoverage in the five years through 1994 (National Council on Compensation Insurance, Inc. [NCCI], 1994).

An important question is whether the media will start to tap into the pool of data from these and other sources to tell us what we can learn about the workers' compensation experience.

"The insurance system can be circumvented by unscrupulous employers to essentially remain uninsured and transfer potential costs to others," declares the *Minnesota Report to the Legislature on Independent Contractors in the Workers' Compensation System,* quoted in Burton (1994). Noting widespread information about uninsured claims from all industries, the report urged a review of the insurance coverage of "thousands of businesses."

"Anecdotal evidence and data available to the Department of Labor and Industry indicate premium fraud may be widespread," the report continued. "The magnitude of the problem is much larger than the amount and structure of resources currently available to address it" (Burton, 1994).

Meanwhile, workers' compensation analyst John Burton has dismissed concerns that worker fraud contributes in a significant way to cost escalation. Instead, he considered the big picture, drawing the issue of employer fraud, where he sees a greater danger, into the loop: "The fraud problem needs to be viewed as part of a broader question, to wit: are we accurately measuring the number of work-related injuries and diseases and the economic consequences of those events?" (Burton, 1994).

What happens when employers hide reportable injuries? We know that it happens. During the 1980s, OSHA began a program to exempt from inspection companies that had good safety records. The agency was forced to abandon the plan a few years later when massive record-keeping fraud was exposed. Bath Iron Works in Bath, Maine, for example, failed to log more than 600 injuries or illnesses that occurred, including serious lung diseases, amputations, and fractures. In another instance, comparing data from two systems, Burton cited a large discrepancy when workers' compensation-recorded deaths for 1991 constituted only 44 percent of the workplace deaths that were listed in a Bureau of Labor Statistics census of occupational fatalities for the same period. Such a gap, he pointed out, has serious economic consequences (Burton, 1994).

It might be added that such skewed figures also have important consequences for public policy decisions and can seriously cripple efforts to make laws and regulations and spend public funds wisely in seeking to protect worker health and safety.

A HISTORICAL BLOC: THE MALINGERER, THE SHYSTER, AND THE QUACK

The impression that there is a crisis in our workers' compensation system is widespread (Rest, Levenstein, & Ellenberger, 1995). Employers are portrayed as besieged by the increasing costs of workers' compensation insurance, while insurers are seen as being forced to "subsidize" the system by state insurance regulators bent on limiting premium increases (to prevent the flight of employers to less costly states; Burton, 1994). The public believes that there is an unholy alliance against American business that is driving the current crisis in workers' compensation. Fraudulent claims by workers ("malingerers")—aided or inspired by ambulance-chasing attorneys ("shysters") and money-grubbing physicians ("quacks")—are seen as responsible for the soaring costs of workers' compensation.

The role of lawyers has received considerable attention in the recent climate of crisis and reform (NCCI, 1994). Often, lawyers are seen as out to make a quick buck in a system that was originally designed to be nonlitigious, without actually rendering their clients any additional financial benefit. Additionally, there is considerable animosity toward lawyers among members of the medical profession. While there has seldom been any love lost between physicians and lawyers, "attorney involvement" is now commonly seen as a factor that will impede the therapeutic relationship and delay a patient's return to normal function. It has become a statistical variable of interest in research (Tait, Chibnall, & Richardson, 1990). Unfortunately, there is little discussion of why workers retain lawyers in the first place, nor of how physicians and lawyers might work together to best meet the legitimate needs of injured workers.

As might be expected, the lawyers employed or retained by insurance carriers and large self-insured employers are not subject to the same impecuniousness as workers' attorneys. Indeed, it would be difficult to even estimate the extraordinary level of legal expenses generated by insurance carriers. For insurance carriers and self-insured employers, top-flight legal advice is a highly prioritized cost of doing business, and few corners are cut in that department. Internally, there may be no real incentive to curb litigation. Payment delays simply make more funds available for investment while frustrating the claimant to the point where (s)he may simply give up or settle for the first offer extended by the carriers. Many injured workers—especially those without attorneys, unions, or adequate financial resources of their own—are thus pitted against overwhelming odds when there is a dispute.

In the recent rash of talk about a national health program, physicians involved in workers' compensation have also come under greater scrutiny. Some apparently charge more for a service delivered under workers' compensation than they would for the same service delivered under other insurance programs (price discrimination), while others label an injury as work-related primarily because they may make more money through that designation (cost-shifting). There is evidence of both types of transgressions (Butler, Johnson, & Baldwin, 1993; Burton, 1994). Some doctors—no easy prey to the blandishments of a quick buck. It is not clear, however, how widespread fraudulent practices are, nor the context in which they occur. Is it possible, for example, that some physicians—seeking to provide the best care for their uninsured or underinsured injured patients—investigate the possibility of work-relatedness more thoroughly in order to provide the care needed. Perhaps they may even employ a less rigorous definition of work-relatedness when their patient needs care that is not otherwise covered. To leap to the conclusion that workers' doctors are cheating without information about context serves the interest of propaganda, not system reform.

Further, the desperate attempts by employers and insurers to limit workers' choice of physicians suggest that they believe they can control compensation costs by controlling or influencing determinations made by physicians. But there is little evidence to support employers' fears about worker choice of provider. Indeed, there is evidence that limiting worker choice of physician may actually *increase* medical costs (Pozzebon, 1993). While employers may have some control over the physicians they actually employ or contract with (i.e., the "company doctors"), truly independent physicians are clearly less subject to employers' influence. Whether or not these physicians are on the endangered species list, especially given the current burgeoning of managed-care alternatives, should be a matter of public concern.

Perhaps most distressing in the discussion of "crookedness" in the workers' compensation system is the issue of underreporting of injuries and disease. Too frequently, "loss prevention" has come to mean "discouraging" workers from filing claims; it has not translated into injury prevention or workplace safety. Some employers have devised creative ways to control their losses. For example, they may discourage injury and claims reporting through so-called safety incentive programs, in which workers receive bonuses or other gifts if they are "accident free" for a certain period of time. (It might be interesting to study medical care cost shifting in this context, especially in firms that provide no health insurance benefits to their employees.) On the other hand, employers may simply have the injured workers report to work and spend their time doing meaningless tasks to avoid "lost-time" indemnity payments. Even "medical only" claims may not make it into the system if the employer simply pays the worker's medical bills instead of forwarding them to the compensation carrier for payment.

That comparatively few cases of occupational disease make it through the compensation system is commonly acknowledged (Barth & Hunt, 1980). This fact is often attributed to various technical factors, including long latency periods; lack of physician recognition of work-relatedness, or the multifactorial nature of many chronic diseases (i.e., the difficulty in identifying the causes of diseases that may have long latency periods or numerous possible origins). These are surely important factors, and they make it relatively easy for insurers and employers to dispute a disease claim. But one wonders about other not-so-technical explanations, as well. As with many work-related injuries, if employers are experience-rated or self-insured, it is surely not in their interest to pay a costly claim readily. Doing so could easily result in a premium increase. Workers may also have their reasons for not filing compensation claims. Aside from the sheer stress of working through the system and having their credibility questioned at every turn, they may find themselves subject to subtle or not so subtle forms of discrimination at work. They may even find themselves suddenly out-of-work.

Employers are required to record occupational injuries and diseases on an OSHA log, and failure to do so has occasionally resulted in large fines by the agency. But similar cheating by employers in connection with appropriate handling of work-related illness and injury has seldom been addressed (or publicized) by state workers' compensation agencies. Built into the structure of workers' compensation are great inducements for employers to misclassify workers' risks, to fail to report injuries and illnesses, and to control "losses" in inappropriate ways. This type of cheating is very seldom a topic of public discussion, although local media are only too eager to report the occasional scandal of the malingering worker enjoying sunshine and recreation in some vacation spot.

THE ORGANIZING OF VICTIMS

One of the major movements of the 1960s—and a harbinger of change in occupational safety and health—was the black lung movement.[4] Disaffected mine workers organized a new group, the Black Lung Association, in part because of the failure of the workers' compensation system to provide help to mine workers disabled by lung disease, in part because of their loss of access to health care when

[4]Victims' organizations and their role in the development of OSHA are discussed more fully in Chapter Five.

they became unemployed, and finally because they felt betrayed by an undemocratic union. The Black Lung Association became a successful trade union group as well as an effective advocate for compensation for mine workers.

In the wake of the successful mine workers' organization, disabled textile workers formed the Brown Lung Association and fought for compensation for cotton mill workers with byssinosis and became important in the struggle for an OSHA cotton dust standard. Similar organizations developed among victims of asbestos exposure, and at one point during the 1970s the Black, Brown, and White Lung organizations formed a "Breath of Life" coalition. These groups turned their illnesses into a way to engage in social struggle. Their members had concrete aims, mostly involving financial compensation for their injuries, but they also challenged their inhumane treatment by employers and insurers, by state agencies insensitive to worker needs, and, in the case of the miners, by undemocratic and corrupt trade union practices.

Sick and disabled workers gained the sympathy of the press in communities hostile to trade unions and won the support of even such right-wing politicians as Senator Jesse Helms, who could hardly deny his victimized constituents their proper due. Sick workers in victims' organizations won sympathy in places where trade unions of well workers were perceived as a threat. There continues to be a coterie of injured-worker organizations that are primarily concerned with the vagaries, injustices and arbitrariness of the workers' compensation system. The success of victims' organizations in the past provides something of a model for the use of political action to improve the current inadequate compensation system in the United States. But truly effective change may require versions of the historic bloc, such as that which was successful in banning the "kiss of death" shuttle in the early part of this century.

CONCLUSION

We have discussed the workers' compensation system in the United States at some length because it brings into stark relief why the point of production is so important to public health in general and to occupational safety and health in particular. The costs—some $60 billion per year—are staggering. The industry that feeds on this system (the insurance companies) are major economic players and powerful institutions in their own right. The ideological issues surrounding workers' compensation are also graphic testimony to the ways in which the media distort public perceptions and fuel an antiworker ideology. The system also engages nearly all major political and economic players: from workers to lawyers, doctors to employers; small business, large business, and the government.

As we have noted, the system developed as the modern state sought to mediate between employers and workers and to reduce the volatility and the social and economic costs of workplace injury and disease. It is a system that penalizes the worker and that provides inadequate compensation for injury. The evidence that it improves working conditions by imposing economic incentives to clean up the workplace is very slim (Victor, 1982; Chelius, 1982). Workers' compensation may encourage some industries to prevent physical injury, but there is hardly any incentive in the system for the prevention of occupational disease.

Prevention should be the backbone of a compensation system. The current system does little to prevent occupational disease and injury, nor does it encourage the internalization of the costs of worker injuries. The compensation system largely externalizes those costs, allowing the employer to shift the burden to the public and to the worker.

CHAPTER SEVEN

The Politics of Occupational Health Science

Because of the imbalance of power in the United States between labor and capital, intermediate strata with responsibilities for worker health and safety and for the general environment have taken on great importance—and the individuals acting in those capacities, if they are committed to serious pursuit of their professions, sometimes find themselves confronted with ethical dilemmas. For instance, the industrial hygienist working for a company that is using antiquated, unsafe equipment is committed to the well-being of the company, and must understand the limits on modernization possibilities imposed by undercapitalization. At the same time, contradictory responsibilities to the health of workers might turn the same person into a corporate whistle-blower.

Often, however, because many professionals are trained to understand health and safety problems as essentially technical issues rather than ones rooted in the social relations of the work-place, the "solutions" they propose may deal only with the most nar-row definitions and marginal issues. The technical staff are trained to solve technical problems, not to redefine an enterprise or recon-struct the mission of a corporation by raising issues such as the ten-sion between the search for profit and worker health and safety.

Power relations, technical training, accountability, and ethics are intertwined in the work of occupational safety and health professionals. But, given labor's comparative weakness relative to management in the United States, finding and winning allies among professionals is a crucial aspect of strategic thinking for improving the work environment.

In a manner consistent with Western neo-Marxian analysis, we have discussed in earlier chapters the regulatory role of the state in occupational safety and health matters as lending legitimacy to existing power relations. The federal-level Occupational Safety and Health Administration, after all, protects workers in the United States from the hazards of the workplace. Where necessary, then, the state has intervened to deal with what is viewed as the essentially marginal problems of a sound, that is, legitimate, system of production. In order for the mass public to believe that the state is acting in its interest, regulatory agencies must have "relative autonomy"; that is, OSHA must be able to take action against bad actors—egregious violators of safety standards—and may even be construed as overly adversarial by some employers if the agency touches vital economic nerves. To the extent that this is so, it provides a window onto the point of production and a way for the state to be used to ameliorate working conditions. The most important actor in the overall role of education, research, and enforcement of health and safety provisions is the occupational health professional, whether he or she works for the federal government, state agencies, or the company.

Within this framework, what is the role of occupational health science? Is it distinguishable from the role of occupational health professionals? Are Gramscian notions of the role of the state under capitalism useful in understanding the relation between science and capital, science and labor?

Take the case of Dr. David Kern, a leading expert in occupational medicine who, until recently, was the director of an occupational health clinic in Pawtucket, Rhode Island, and a professor at Brown University. Here is his story as reported by a Providence newspaper:

THE MYSTERY OF OCCUPATIONAL DISEASE

PAWTUCKET—Memorial Hospital has fired a Brown University professor who defied hospital officials by presenting a scientific paper in May about his discovery of a possible new lung disease.

Dr. David H. Kern, who has worked at the hospital as a medical researcher since 1990, has been notified that his five-year contract, which will expire in two years, will not be renewed. Although Kern will still be allowed to work at Memorial for two years, hospital officials have told him his occupational-medicine program will be discontinued. Kern says he intends to leave.

Kern said such nonrenewals occur only under incredibly unusual circumstances and that he is being "punished" for a dispute with the hospital and with a Pawtucket manufacturing company, where he found an abnormally high number of workers with serious breathing ailments.

Kern asserted that Memorial bowed to pressure from the company, Microfibres Inc., to keep him from releasing his findings about the mysterious lung disease that eight employees or former employees had contracted.

"The real issues continue to be that a company and a hospital acted together to prevent the dissemination of scientific findings that have public-health importance," Kern said yesterday. "They did so because of economic considerations that should not be allowed to see the light of day."

Kern also said that Brown University, whose medical school has an affiliation with the hospital, has compromised the principles of academic freedom for not standing behind him during his conflict with the hospital.

The dispute arose from Kern's investigation of Microfibres, which makes polyester, nylon, rayon, acrylic fibres and velvet fabrics for upholstery. He went to Microfibres in November 1994 with a group of medical students as part of a training program that required him to bring students in the occupational-medicine program to a workplace every six weeks.

Kern was interested in Microfibres because a lung specialist referred to him a patient who worked at the company and who was suffering from a shortness of breath and had a chronic cough. The patient had an uncommon lung ailment called interstitial lung disease, in which lung tissue becomes inflamed and scarred, making breathing difficult. The specialist believed his ailments were work-related. Kern and his students signed agreements at Microfibres in which they promised not to reveal trade secrets.

Kern said his initial trip and investigation found nothing conclusive to link the patient's symptoms to Microfibres. But more than a year later, in January 1996, Kern came across another Microfibres worker who had similar symptoms. Kern contacted company officials and suggested that they notify the National Institute for Occupational Safety and Health (NIOSH), which they did. Kern subsequently found five more cases of a similar ailment at a Canadian plant that is owned by Microfibres.

Microfibres hired Memorial Hospital—and thus Kern—to investigate the matter, and also was working with NIOSH and other consultants. Kern eventually turned up six additional cases of interstitial lung disease among 150 employees in the Pawtucket plant. Eight cases is much more than would normally occur in a group of 150 people; typically there is 1 case of the disease per 40,000. But the company and Kern continually clashed over how the study was to be managed and what the company's employees would be told. Kern said that Microfibres would not allow him access to air-sampling data compiled by NIOSH. Earlier this year, hospital spokesman Rick Dietz saw it this way: "The working relationship with Dr. Kern and the company was not strong. The way they wanted to manage the investigation and the communications was just different. They could not find a common ground to make it work."

In October 1996, Kern gave Microfibres a summary of his findings, and told the company he was planning to present it to a meeting of the American Thoracic Society. The company urged him not to and said it would sue him if he did, Kern said. Kern consulted with Brown University, and the university advised him not to submit it, citing the confidentiality agreement that Kern had signed. Then in December, Francis Dietz, the hospital president, wrote a letter to Kern asking him not to present his findings at the meeting. In the next paragraph, Francis Dietz told Kern that Memorial would be closing its occupational-medicine program. Essentially, that meant Kern would continue to be a hospital employee and see patients, but he could not continue his work with industry or train Brown students in occupational-health issues.

Kern went ahead and presented his research to a meeting of the American Thoracic Society in San Francisco in mid-May because he thought the material needed to be presented to the scientific community. About two weeks later, the hospital notified Kern that his contract would not be renewed. The notification came just two days before the hospital's deadline of giving notice that his contract wouldn't be renewed.

Rick Dietz, a spokesman for Memorial Hospital, could not be reached yesterday for comment. However, he told the Boston Globe that Kern's contract's being terminated "has very little to do, if anything," with the ongoing dispute over Kern's right to release his findings. "It has a lot to do with Dr. Kern's interests in the long run, the hospital's needs and the ability of both parties to be working in a common direction."

In an interview earlier this year, Rick Dietz said Kern was never asked to suppress information. "He was asked to postpone it. There was never a demand not to submit. We're not going to stand in the way of academic freedom."

Kern asserts there was economic pressure put on the hospital to squelch his findings. Microfibres has acknowledged that Memorial did approach the company during the hospital's fundraising campaign, seeking a donation for a new Center for Primary Care that Brown wants to establish at the Pawtucket hospital.

That solicitation occurred "well after" Kern's work on the case ended, Microfibres said in a statement. The company said it did not make a donation in response to that solicitation, but "the company has had an active program of charitable giving in the community over the years and intends to continue it." The proposed Center for Primary Care "is the type of community project Microfibres would support."

Kern's plight has not gone unnoticed by his colleagues at Brown and by other occupational-health experts around the country. They have written the university some 100 letters of support. Kern also has the support of James Celenza, director of the Rhode Island Committee on Occupational Safety and Health. "Aren't we—as doctors, as health professionals—obligated to make known the potential health threat?" Celenza wrote in April in a letter to Donald J. Marsh, dean of Brown's Medical School. "If Dr. Kern had agreed with both the hospital and the medical school, how would other health professionals or workers learn about the potential lung problems associated with this situation?"

Brown University insists that it has been supportive of Kern and has not compromised academic freedoms. The university has said that it merely advised Kern not to make a presentation because the legal ramifications of the confidentiality agreement he signed were unclear, according to Mark Nickel, a university spokesman. "We absolutely support the right of faculty to research and publish," he said. "The university faculty rules prohibit doing any classified research for which you are unable to share the results freely."

To date, Microfibres hasn't sued Kern. But David Monti, who works for Fern/Hanaway Monti & Partners, a Warwick advertising and public relations firm representing Microfibres, said the "door is still open." Monti could not say whether Kern's address in San Francisco in May revealed any of Microfibres's trade secrets. Microfibres continues to work with other consultants and occupational health experts to pinpoint the source of its employees' ailments, which still remains a mystery, he said.

As for Kern, he's not sure where he'll go from here. "I don't, at this point, have any intention of staying," he said. "When I leave remains unclear. . . . In that environment I can't function in a meaningful way."

The loss of Kern's program at Memorial is also a loss for the state. It is the only occupational health research and clinical care program for injured workers in Rhode Island. (Barmann, 1997)

This chapter explores the social and political dynamics of the occupational disease "mystery."

THE SOCIAL LOCATION OF OCCUPATIONAL HEALTH PROFESSIONALS

Work environment "professionals" are practitioners and include ergonomists, industrial hygienists, occupational physicians, occupational health nurses, safety engineers, ventilation engineers, trainers, and other occupational safety and health specialists. By way of contrast, "scientists" are researchers and include a wide range of medical and engineering researchers, as well as epidemiologists, toxicologists, physiologists, chemists, industrial sociologists and psychologists, as well as researchers trained in the "professional" fields.

Thus, professionals and scientists in occupational health perform related but different functions. Scientists of various disciplines seek to develop basic knowledge about the hazards of the workplace. Professionals apply scientific knowledge to solve problems; they are technologists. In the past, professionals were "independent"—they were governed by professional ethics but worked for clients. One of the features of advanced capitalism is that many of these independent professionals have become "employees," their positions either made internal to large corporate bureaucracies or they are consolidated into consulting firms.

Understanding the professional role requires identifying the social location of the professional: professionals themselves often are substantially and directly under the control of employers or, via the market, of powerful clients. Ethical and professional standards usually include obligations to the public, which are perhaps constraints. Many professional associations have highly developed codes of conduct. Despite the existence of such ethical constraints, professionals remain primarily under the control of employers. On the other hand, they are "workers"—that is, they do not own the firm. Thus, there may be a natural tension between their obligations to the employer (essentially to supervise worker health and safety) and their identification with workers (as fellow workers and advocates of health and safety).

For example: a few years ago, Violet, a young industrial hygiene student who had no work experience in the field, was lucky enough to be given a summer internship with a petrochemical company.[1] The student had an undergraduate degree in biology and had become interested in occupational health through her concerns about the environment and sympathy for workers and trade unions she had learned in her "progressive" family. She was nervous about working for the company but was very pleased when young staff who had graduated from the institution at which she was studying welcomed her.

After a half-day's orientation by management, she was issued a white hardhat and introduced to an older worker. Her supervisor asked him to show her around the chemical plant, indicating that she would be with them for the summer. Her guide wore a blue hardhat (as did all the hourly workers).

Violet was then given a truly grueling tour of the petrochemical plant, climbing ladder after ladder to the tops of tank after tank. Subsequently, throughout the summer, she was teased unmercifully, was told tall tales, and was sent on fool's errands. The workers assumed that she was a management trainee, probably an industrial engineer of some sort, and treated her accordingly. By the end of the summer, this environmental activist hated workers. Not until the fall did she understand that her treatment had been mostly a matter of wearing the "wrong-colored" hat in a situation where there were only two colors.

THE POLITICS OF PROFESSIONALISM
AND THE STATE

In light of this overwhelming economic imperative, which was unquestioned among those who controlled advancement within the industry, the mining engineer was compelled to subordinate whatever

---

[1]"Violet" is a composite character constructed from the experience of several industrial hygiene students (none named "Violet"), as reported to author Levenstein.

technical considerations might create conflicts to more comprehensive and commanding economic considerations. The social structure of the industry prevented him from making common cause with the miners who suffered the consequences of unsafe and unhealthy working conditions. . . . But it would be wrong to presume that many mining engineers found this to be a particularly stressful situation. . . . When asked why he did not provide more roof supports to prevent rock falls that routinely killed his immigrant miners, an operator is said to have replied that "wops are cheaper than props." (Donovan, 1988, p. 100).

In Ronald Bayer's collection of case studies in ethics and the history of occupational health, the tendency of work environment professionals to act as relatively low-level members of management is made clear. "The history of asbestos regulation gives sorry confirmation of the adage that what you see depends on where you sit. And, one might add, what you say has much to do with who owns your chair" (Murray, 1988). The puppetike behavior that is depicted in these histories confirms our own understanding of the political economy of occupational health and documents a social structure in which worker health consistently pales in importance, compared with the economic imperatives of industry. But this exposition does not treat the problem of agency—that is, who or what accounts for any improvement in working conditions that occurs when labor is weak and professionals are servants of capital.

Modern capitalism's overwhelming hegemony in matters economic seems to leave scant room for reform. Yet, specific reforms have occurred, and, with the passage of the Occupational Safety and Health Act (OSHAct) in 1970, a new arena for struggle around working conditions was created and a new type of occupational health professional came on to the scene. In many ways, the 1960s were a turning point for the United States, and the emergence of a new breed of middle-class intellectuals had important repercussions for occupational health.

THE OLD AND NEW LEFT

Doctors, nurses, and other health professionals have constituted, since the late nineteenth century, a small but active cadre of leftist-oriented public health workers. For example, a report published in

1891 by socialist Elizabeth Morgan and her comrades in the Chicago Trade and Labor Assembly led to the passage of the Illinois Factory Inspection Act (Lear, 1992, pp. 301–302). At the turn of the century the activities of the groups around Hull House and the work of Alice Hamilton, Crystal Eastman, and others paved the way for labor legislation on the conditions of work and for the creation of the Department of Labor, in 1913. The first socialist organization of health workers was the Dentists Study Chapter of the Intercollegiate Socialist Society (which was a precursor of the League for Industrial Democracy). This institution also published the first avowedly leftist health worker journal, *The Progressive Dentist* (Lear, 1992). The first Worker's Health Bureau was established in 1921 by women with experience in the labor and consumer movements. It was short-lived (being disbanded in 1928), but while in existence it organized the first national labor conference on health, in 1927 (Rosner & Markowitz, 1984). In the 1930s Local 1199, a union of New York City pharmacists and drugstore clerks, was formed under Communist party leadership, raising the issue of the role of middle- and lower-middle-class professionals in the labor movement. Local 1199 was later to become the National Union of Hospital and Health Care Employees. In addition to being involved in the founding of Local 1199, progressive physicians also formed the Physicians Forum, which did pioneering work on racial discrimination and, later, on the health effects of atomic weapons; this latter concern, in turn, led to the development of Physicians for Social Responsibility (Lear, 1992, pp. 304–305).

The old romance between intellectuals and the labor movement had soured in the 1950s. On the one hand, the young trade unions of the 1930s needed the skills of the intellectuals, and the intellectuals' education gave them power in the labor movement. On the other hand, the rise of the Cold War and McCarthyism resulted in the purging of the Left—and many left-wing intellectuals—from the dominant trade unions. By the 1950s, the relatively mature trade unions could hire economists and other essential technicians, and elected officials had learned to wield and protect their own authority. In any case, the romance may have been one-sided. American anti-intellectualism reasserted itself, and the Left for its part, became disenchanted with the "new men of power," their business unionism, and their limited social vision.

In the 1950s science and technology achieved its broadest currency as the underlying religion of capitalism and socialism. Nuclear weapons and nuclear power, competition in outer space exploration, rapid industrialization in the Soviet bloc, and the rise of consumerism in the West all signaled obeisance to the hegemony of scientific thought—and brought about widespread scientific education for the emerging middle classes. By the 1960s, the rise of "New Left" intellectuals included on their part not only a commitment to the classic egalitarian principles of the democratic Left but also a critique of, first, the uses of technology and, later, of technological thinking itself.

Another thread in the development of a leftist-oriented public health movement was the creation, in 1964, of the Medical Committee for Human Rights (MCHR). Deriving from the activities of New Leftists and liberals, in particular, civil rights activists, MCHR moved from the liberal left and became increasingly radicalized. MCHR activists formed chapters throughout the United States, opposed the Vietnam war, and were active in providing medical care at demonstrations. Many members of MCHR had also been involved with the New Left's free clinic movement.

MCHR activists played a critical role in two especially relevant areas. The Chicago chapter developed the first occupational safety and health project, which later became the Committee on Occupational Safety and Health (CACOSH). Founded in 1972, the CACOSH, was the first of what was to become a national movement of COSH groups.

Similarly, MCHR members linked up with union activists in Philadelphia—particularly Anthony Mazzochi of the Oil, Chemical, and Atomic Workers [OCAW], who had played a pivotal role in organizing labor to support the 1970 enactment of the OSHAct and who was a major player in occupational and environmental struggles throughout the 1970s and 1980s—to form the Philadelphia Area Project on Occupational Safety and Health (PHILAPOSH).

In Massachusetts, a small group of activists at Urban Planning Aid (UPA—a federally funded institution and organizing group that had its origins during the Johnson administration's War on Poverty) created a job safety and health project in the early 1970s. This project led to links with Local 201 of the International Electrical Workers at the large General Electric plant in Lynn, near Boston. The safety committee that the union and UPA people created provided the focus for struggles around the health and safety issue, and, when federal aid to UPA was threatened, this group helped put together the Massachusetts Coalition for Occupational Safety and Health (MassCOSH), established in 1976 (Berman, 1981).

MCHR activists also linked up with Ralph Nader's Public Citizen Group to create the Health Policy Advisory Center and a journal, *Health/PAC,* that has played a pivotal role in disseminating a progressive public health perspective. The Nader group also published *Bitter Wages,* a major book on occupational safety and health issues (Page & O'Brien, 1973).

Similar COSH groups sprang up in California, New York, Pennsylvania, South Carolina, North Carolina, Connecticut, Rhode Island, and elsewhere. During the mid- to late 1970s COSH groups struggled to improve the protections offered by OSHA and to push for greater visibility for the issue. In this effort they were aided by a cadre of professionals developing around the new interest in health and safety and prompted by the existence of OSHA itself. Indeed, for a brief period during the late 1970s COSHes benefited directly from OSHA activities. The appointment of Eula Bingham (a progressive academic and industrial hygiene expert) as head of OSHA in 1977 promised a more active regulatory agency, and her creation of OSHA's New Directions program provided an important source of funding for a number of COSH groups and related organizations.

Most COSHes built upon specific links between health professionals, former New Leftists, and trade unions. These groups brought to the issue of occupational safety and health a unique perspective: they linked middle-class professionals, radicals, and around a reform issue. COSH groups stressed that improvements in the work environment must originate in workers' knowledge and experience about workplace hazards. But using this knowledge required access to technical information and a political philosophy based on worker empowerment (Rosenstock, 1992). By reaching out to progressive technical experts and functioning as worker advocacy organizations linked to the labor movement, the COSH movement did much to reconnect the rank-and-file union member with progressive intellectuals and technical specialists. In this respect, the ideology of the COSH movement connects the Old Left's traditional reliance on workers and unions with the New Left's focus on empowerment, control, and the use of technical knowledge.

Frequently, these new organizations were created with a heavy dose of skepticism or hostility by the official labor movement in the area. Today, however, COSH groups function in some twenty-five states and have achieved recognition from the national AFL-CIO as a significant force for workplace reform. The head of the AFL-CIO health and safety department is an alumna of MassCOSH, as is the head of the United Steelworkers health and safety department. The insurgents of the 1960s and 1970s have become labor's mainstream health professionals of the 1990s.

THE NEW PROFESSIONALS

As already suggested, the passage of the OSHAct in 1970 created a new arena for potentially improving working conditions. It also provided many opportunities for the employment of occupational safety professionals, not only in the agency itself but also in the "regulated community," in consulting companies and insurance carriers, and in national labor unions. The OSHAct did not specifically empower workers, but it did provide an expanded role for health-related professionals and did create a new platform for critical discussion of science and public health. Each attempt by OSHA to set a new standard in effect invited labor's and management's health professionals to counterpose their best arguments and evidence in a highly visible national arena. At local and regional levels, the new standards created new demands and opportunities for medical and technical services. The "new sheriff in town" was not a deputized worker-inspector, or a even a joint committee, but the former summer intern at the petrochemical company, whose biggest mistake that summer was to wear the "wrong-colored" hardhat.

Perhaps most prominent in stimulating the genesis of newly trained professionals were the Educational Resource Centers (ERCs), funded by the OSHAct-created National Institute for Occupational Safety and Health (NIOSH). Regional centers around the country provided subsidized training at the masters and doctoral levels for industrial hygienists, safety specialists, occupational physicians, and nurses. The NIOSH funding for such centers and for extramural academic research permitted the development of professionals relatively independent of industry. In addition, continuing education programs sponsored by the NIOSH ERCs helped to elevate the skill levels of practitioners. For the first time, significant resources were being devoted to creating a professional base for occupational safety and health in the United States.

By contrast, labor's comparative strength in the United Kingdom resulted in substantially different legislation, mandating a national tripartite Health and Safety Executive and giving workers and their representatives a much greater role in standards development and their enforcement. There was no accelerated development of professionals, however, and as a consequence the Conservative Thatcher government was able to undercut workplace safety regulation by significantly reducing the economic and political power the British trade unions. Tripartism cannot work if one of the "parties" has been decimated. In the United States, the more highly developed health and safety bureaucracy, although battered, was able to survive the equally conservative Reagan administration (Wooding, 1990).

In Scandinavia, with its highly organized labor force, we see the limits of professionalism. National occupational health and safety legislation in Norway, Sweden, and Denmark did empower health and safety committees and worker-inspectors, even in some instances making professionals accountable to labor–management joint committees with a worker majority. Nevertheless, critics of the Scandinavian system argue that the system has have not been able to make substantial improvements in injury and illness rates because of the narrow definitions that professionals have imposed on problems, and which the worker committees and trade unions have largely accepted. Some have argued that only major reform of work organization will truly improve the health of workers and that the industrial hygiene/expert-dependent model will not allow for such change.

Karasek and Theorel also argue for a more holistic approach to the work environment, as do numerous "participatory action researchers" in Canada, the United States, and western Europe (Karasek & Thorel, 1990). These critics suggest that the relationship between experts and workers must be radically altered in order to create a more humane workplace.

There is little evidence that the contending approaches whether expert-based or worker-empowered in their primary orientation, are substantially better or worse than one another in improving health and safety on the job. One would suppose that various approaches may be more or less suitable in different situations. But also, some critiques operate at an ideological level and so may have little to do with actual health and safety effects. Is "the problem" essentially "technical" or "political"? Must one be an expert to operate effectively in this arena, or can activists or trade unionists that with only minimal technical training also be effective? While proponents of the nontechnical, essentially political nature of the problem have primarily been leftist intellectuals (though professionals in their own right), some trade unions have asserted political control of health and safety training and promoted the use of "peers" or worker-trainers with a barebones trade union ideology (Wooding, Levenstein, & Rosenberg, 1997). Other unions have employed the hired-hand approach, using technically trained depoliticized experts.

Inevitably, there are tensions between trade unionists and professionals. Because labor and management already contest the terrain in which both groups operate, trade unionists wonder whether the professionals will simply ally themselves with the employer who most frequently pays them, or will their affirmative responsibilities for worker health embolden them to be—at the least—truly objective? Will the professionals employed by the state be any more independent of the employer, or will they be driven by the imperatives of the marketplace, though in more subtle ways than for the employers' professionals? Will the proworker professional who focuses on health and safety be at peace with the trade union leadership (and membership) who have a broader agenda? Will the professionals in advocacy groups be so proprietary of and dependent on a special relationship their group has with a trade union that they end up undercutting the role of state regulatory agencies? Who will "own" the work environment—the employer who owns the property, the trade union whose members want to control the labor process (or at least get paid well for giving up control), the COSH group that advocates for dissemination of health and safety information to workers, the government compliance officer, or the corporate staff hygienist? The possibilities for conflict are enormous.

In much of eastern and central Europe, company personnel with responsibilities for health and safety are members of the same union as the production workers. In the United States, the more narrowly defined labor movement needs political and economic allies in its efforts to control workplace technology and its attendant hazards. COSH groups represented an interesting and unique way to bring together labor and public health experts. However, such alliances are fragile, as they are built on conflicting allegiances, differing definitions of the problem to be addressed, and occasionally inflated self-images of the participants. If the creation of coalitions were an aspect of redefining the labor force and the labor movement, perhaps the tasks ahead would be easier.

The current political and economic climate is quite different from that of the 1960s and 1970s, when the national structure of occupational health and safety was created. Since the beginning of the 1980s, the state has been in headlong retreat, and the ideology of the market has run rampant. These trends have eroded the role of the professionals and of the health and safety bureaucracy essential to the American national model. Will labor mobilize to defend the embattled work environment? Can a revitalized AFL-CIO address the massive problem of organizing American workers? Can health and safety concerns play some role in such an effort? While there is some recognition of the deteriorating condition of the American working class, we do not yet have any indication that trade unions are willing to embark on dramatic new campaigns to turn the situation around. There is no significant left in the United States and there is a very significant right. The antiregulatory antigovernment ideology that characterizes much of American political discourse leaves scant room for an independent progressive middle-class and professional movement; the ball is in labor's court.

SCIENTIFIC RESEARCH AND THE PRIVATE SECTOR

Fundamental to the ability to control or eliminate occupational health and safety hazards is the recognition of their existence. The "right to know" is meaningless if there is no information available on hazards. Sometimes it is obvious: an acute traumatic on-the-job injury is difficult to miss. Nevertheless, a substantial number of such occurrences never make it into the official occupational health statistics. Reacting to these shortcomings, NIOSH has been supporting state-level investigation and registry systems in order to get a full accounting of the human health toll of workplace technology. In Massachusetts, neither the workers' compensation system nor the OSHA data collection system has captured even the majority of acute traumatic fatalities.

The traditional Left/syndicalist response to this problem has been to focus attention on workers' own knowledge of workplace injuries and disease (Wegman et al., 1975). However, the short-term effects of chemical exposures may not always be obvious—even in cases where the long-term effects are conclusively fatal. In addition, the deep mistrust of worker reports by employers is often echoed by science, in that "objective" evidence and analysis are more highly valued and weighted than the "subjective" signs and symptoms of those allegedly injured.

It is impossible to measure with any degree of confidence the amount or origin of resources devoted to occupational health and safety research in the United States. The government funds most of the research reported in the leading peer-reviewed journals, although it is clear that an enormous amount of scientific investigation occurs in industry apart from the involvement of government. Plainly, the budget of NIOSH does not represent the full extent of research in this field; in fact, government scientists are reluctant to identify all of the various government sources of worker health and safety research, for fear of attracting the attention of antilabor budget cutters. The various national institutes of health, the Department of Energy, the Environmental Protection Agency, the Department of Agriculture, the armed services—all are interested in worker health and safety.

Scientific research in occupational safety and health may be conducted by academics, consulting companies, research institutes, insurance companies, and, in some large firms, in-house research operations. Sources of funding include the companies themselves— ones that may suspect they have an occupational health problem or that are in the business of providing technical solutions to such problems. Other sources include government agencies, foundations, industry associations, industry-related institutes, joint labor/management institutions or committees, and occasionally trade unions themselves. No matter what the funding source, however, industrial health research necessarily involves scientific researchers gaining access to privately owned property and/or information and processes.

Financial support for research may come as a grant, a contract, or some type of consultancy arrangement. Grants generally give the researcher the greatest autonomy, with generally few strings attached by the provider. In the case of grants, funds are provided for the investigation of a particular problem, although occasionally whole programs or centers are funded with a more general mandate. Contracts involve deliverables, that is, the researcher agrees to provide a report on a particular problem within a certain period of time. Consultancies can involve any of a wide variety of arrangements, some similar to contract research, others merely agreements to provide advice or investigation as requested. Perhaps the most critical question comes down to: who owns the research product? Frequently, research done under contract or through consultancy arrangements with industry is owned by the industry paying for the work. Even research done under grants may require clearing reports of findings with the funder, although some universities frown on such a practice.

While the significance of ownership of such intellectual property is obviously of great concern, another more subtle but sometimes quite important issue is, Who poses the question to be investigated?Although, most frequently, worker complaints initiate the discussion of issues that eventually become research problems, the actual research question may be framed by the company, by joint labor/management discussion (at either the firm or industry level), by industry associations or their research institutes, or by government through its regulatory policy process or already mandated surveillance. On occasion, scientific advisers or the researchers themselves may be given more latitude in framing the problem. It is rare that labor organizations themselves organize research on health and safety complaints.

The various parties do frame research questions differently. Workers have a fairly straightforward question, namely, Is my work making me sick? Management would like to know whether, in fact, the worker is sick and, second, whether the illness is attributable to the work environment and likely to result in financial liability. Scientists ask whether a specific, measurable health effect is associated with specific, measurable exposures. It is miraculous when the results of any research project satisfy all of the interested groups. Thus, the careful framing of the question and the control of the interpretation of the answer both become matters subject to intellectual property rights.

This returns us to a basic, less subtle, question: who decides *who* should get the research dollars? NIOSH has developed a set of research priorities and encourages external nongovernment researcher involvement in such efforts. In addition, it manages a competition for research support that includes peer review and priority scoring of applications. Insurance companies normally provide in-house services to clients, occasionally fund a limited number of academic research programs, and sometimes employ consultants.

While companies and unions also hire consultants, the lion's share of resources in this society are unquestionably in the hands and under the control of corporate management. Labor is important in making complaints, not in disbursing funds for investigation. Leading academic investigators in the field generally understand that private support from industry is normally an essential aspect of building and maintaining a research program.

THE RESEARCH CONTRACT

Among the issues that emerge from the foregoing examination of the relationship between industry and academics is the two groups' different understanding of the purpose of research, the proper methods of conducting research, and of the responsibilities concerning dissemination of research findings (Quinn, Levenstein, & Rest, 1996). Academic investigators are steeped in a culture committed to scientific discovery, to finding out "new" things.

Research methods are designed, redesigned, and refined during the course of a project in a way that is intended to protect the scientific validity of the work but that will also lead to new and, hopefully, important findings. Once the research has been completed, scientists are bound by their training and by professional ethics to publish results in appropriate peer-reviewed journals and/or books. In the course of the conduct of investigations, scientists may issue preliminary, more speculative, reports as a way of eliciting discussion and criticism from the scientific community, and in the case of health scientists, perhaps issuing preliminary alerts.

By way of contrast, private companies are preoccupied with private property—including intellectual property—and related financial costs and benefits. This should not be a surprise, since the pursuit of profit is what we ask of firms in a market economy, but frequently it is a surprise to scientists imbued with a different ethic. In fact, within a firm, there may be quite different points of view concerning research; that is, the company's own scientists may share much of the orientation of the academic researcher, while corporate lawyers will be on guard to defend the company and its assets.

Essentially, concerning occupational health research, firms seek the service of scientists to be reassured that they are not courting liability through the use of unsafe production processes. Alternatively, on occasion, they are seeking out answers to clearly defined problems, but certainly they are *never* not interested in what they might view as a "fishing expedition." The firm thinks of contracts as legally binding instruments, setting up exact terms of a relationship and limiting the freedom of the researcher to explore. The academic scientist, on the other hand, may view a research contract as a license to explore.

SCIENTIFIC ADVISORY COMMITTEES

In a world in which science has become highly politicized and in which economic stakes may be high, industry may rely on academic and/or scientific advisory boards to provide current information and assistance on occupational and environmental health and safety problems. An important function of a scientific advisory board is (also) to legitimize scientific investigations conducted either under the aegis of private industry or with its resources. The public—or the scientific community itself—may be suspicious of company-sponsored research on worker health without an apparent guarantee of objectivity of the research. A scientific advisory board may set research agendas, design requests for proposals, review the proposals (as a "peer-review" group) and recommend or decide on awards, monitor ongoing industry-funded research, and review research findings. A scientific review board may decide or make suggestions concerning the communication of findings to workers and the public.

The role of such boards can be controversial. The responsibilities and authority of the board may not be clearly thought out or stated, and misunderstanding may result. The following is an extended example of how an advisory committee may not necessarily provide a depoliticized solution to a research issue. An industry association was concerned about the possibility of reproductive hazards from chemicals commonly used by its members, decided to sponsor a "definitive" study with a multi-million-dollar budget contributed by member companies. An advisory board was assembled of leading academic researchers, and interested research groups mandated the board to issue a request for proposals, the first step of which was a call for the presentation of credentials. Groups whose presentation of qualifications was deemed acceptable by the advisory board were then asked to present proposals on how they would investigate the issue. Not all of the groups responded to this invitation (at least one investigator voicing the view that the process was "wired"), but the majority did submit proposals. The advisory committee reviewed the proposals and selected the three most promising, asking the lead investigators to clarify some aspects of their methodologies. One university's proposal occupied a clear first place according to committee members, before and even after the clarifications were made. The committee then reported to the industry association the three (unranked) acceptable proposals—and the industry group rapidly proceeded to choose the least meritorious proposal of the three. The lead researcher of the winning group was well known to the industry participants because of previous research.

Where was labor's involvement in this process? The industry was, generally speaking, not unionized, but a coalition of labor and environmental groups was monitoring the process. When the advisory board went through its first review of proposals, the identities of the leading groups quickly became common knowledge. The group that was initially in first place was well known to be friendly to labor and committed to worker health and safety. An associate of that group had ever been approached by a labor-support organization and told that the reason that the issue of reproductive hazards was on the table at all was because of labor pressure; therefore, labor wanted some sort of advisory role in the study. The request was bluntly put: the academic investigators "owed" something to the coalition of environmentalists and labor groups. Nevertheless, when the first choice of the academic advisory committee was *not* chosen by the industry association, the labor support groups kicked up no public fuss. Advocates that were normally excellent at gaining the attention of *The New York Times* for issues with which they were concerned refused to take the industry action to the press. Rather, they attempted to convince members of the advisory committee that they should have a monitoring role in the research process. Needless to say, the failure of labor to actually deliver for its "friends" in the university left a sour taste in the mouths of the academics—and a lesson learned, perhaps, that such labor activists were not very effective allies.

By way of contrast, labor and management in the auto industry have created special funds through collective bargaining to support occupational safety and health research. A scientific advisory board includes labor and management scientific representatives, as well as academic scientists chosen by the two parties. Certainly there is politics operating within the committee, but it is not subterranean jockeying—and the politics is not a facade for unilateral decision making by industry. In part, this is because the United Automobile Workers remains relatively strong and, additionally, its position vis-à-vis the industry is deeply institutionalized.

WORKER RIGHTS
IN OCCUPATIONAL HEALTH RESEARCH

The relative autonomy of scientific research in occupational and environmental health rests on its legitimizing function, its objective stance, so that its findings are accepted in general—and in particular by workers and citizens who may have reservations about technology. The scientific stance, skeptical of all that is subjective, relies on objective measurement. Thus, no matter how many times researchers validate worker perception of hazards, science remains wary of such self-reporting. No matter that the product of "objective" measurement is frequently further uncertainty, the researcher is cautioned against "soft" science, qualitative measurement, "quasi-science," and relies on the objective stance for status in profession and society.

We are not quarreling with the importance of advances in our understanding of human health and environment that may result from these methods; rather, we are concerned with the vast amounts of experience and data that are thrown out, uninvestigated and unanalyzed, because of the narrow views of legitimacy in this field. And we are concerned that the subject of the research, those with the greatest stake in the outcomes, those with the most unambiguous interest in finding out the truth, the workers, are literally objectified and sometimes humiliated in this process.

Thus, we are presented with a profound contradiction. Workers need allies in their struggles for health and safety. The most obvious allies that they can have are the public health scientists whose responsibility is to investigate occupational safety and health hazards. However, scientists frequently mistrust the workers' own reports of signs and symptoms whereas workers may experience the scientists as cold, analytical, and inhumane. Workers must depend on the scientists for confirmation, validation of common knowledge, and for discovery of the subtle, insidious threats to their health. Nevertheless, workers often have deep misgivings about looks scientists—and the scientists return the insult. There are certainly scientists who struggle with the contradiction between their legitimizing role and their mode of scientific inquiry. "Barefoot" epidemiology, participant action research, and labor and community approaches to risk mapping, for example, are all methods that have been devised to try to overcome these contradictions, just as the Freirian approaches to worker health and safety training have been efforts to put expertise in the *service* of workers rather than in their oppression.

Workers in the United States have a long history of anti-intellectualism, yet they participate in a culture that empowers experts at the expense of citizens. The low value put on hard work and the blue-collar worker in the United States contribute to the difficulties of negotiating a reasonable relationship between science and labor.

Additionally, occupational health and safety research frequently involves human subjects and there are ethical issues in such research. There is the question of "informed consent" at the beginning of a study and "high-risk worker notification" when hazards are discovered after exposure. The National Institutes of Health (NIH) has elaborated procedures for both government research and government-funded research. Virtually every university has an institutional review board, ostensibly to protect the rights of human subjects but perhaps even more importantly to avoid financial liability from mistreatment of human subjects. These guidelines and procedures, however, do not quite ring true in the workplace. Worker-subjects of investigation are different from "patients" in some fundamental way. What does "consent" mean in the workplace? What does "informed" mean?

There are many discussions about what constitutes effective, meaningful communication about risk. But the issue of consent in the workplace has not been explored, to our knowledge. Presumably, when we say "consent," we mean that it is freely given, and the following issues need to be addressed.

1. Can workers be required to participate in a study as a condition of employment? If workers "consent" under those conditions, isn't the level of coercion substantial enough to warrant ethical concern?

2. If workers provide personal medical information to the employer as a condition of employment, does that imply "consent" to use of that information for research purposes? If workers are required to provide health information as part of a preemployment screening, does provision of the information imply "consent"? If medical monitoring is required by OSHA or other government standards, can such data be considered as freely offered for research purposes by the workers?

3. If workers refuse consent, for whatever reason, does the researcher have obligations to protect the workers from related employer discipline? What could a researcher be reasonably expected to do?

4. If the employees' union is not party to the research contract (or not involved in some meaningful way), does that change the way in which we view worker "consent"?

5. What sort of assurance should the researcher require or ask of the employer regarding these matters?

At the other end of the research process is the dissemination of research findings, which may range from publication in journals to letters to high-risk workers and reports to their employers. On the other hand, it may involve no communication of results at all other than to the employer, who paid for the research. For academic scientists, the right to publish findings is critical. Such questions arise as whether the sponsor has rights to review drafts or abstracts. How much time will be given to the sponsor to review the document, and does the sponsor have a right to suggest—or even insist on—editorial changes? In whose hands does the ultimate decision to publish rest? Again, all the questions of intellectual property rights come to the fore. The leading elite universities have strict regulations about only accepting grants and contracts that preserve the academic freedom of the researcher. On the other hand, at one leading university, a researcher seeking corporate funds that were under the control of the dean of the school of public health was asked to assure that his findings (though yet to be found!) not embarrass the dean with the donor. In another situation, an abstract submitted by a researcher had to be withdrawn from an international conference because the industrial association reviewing the abstract as a condition of the research contract objected to speculation about the potentially carcinogenic properties of the substance being investigated.

The power of the corporations to withhold donations, fail to renew contracts, even blackball researchers so that they cannot gain access to workplaces has a "chilling effect" on research and researchers. Our own review of leading occupational health journals shows that the great majority of peer-reviewed scientific articles are funded by the public sector.

What rights to information should workers involved in research projects have? When the labor movement sought to get federal legislation passed to mandate and provide funding for dissemination of research results to high-risk workers in NIOSH industrywide studies, the effort failed. Apparently industry was concerned that workers would take this information to court to gain compensation for their industrial illnesses, which had previously gone unrecognized by the workers' compensation system. Alternatively, perhaps even more dangerous to business interests, there was the possibility that residents near industrial facilities would make the obvious connection between workplace exposures and community environmental problems. As pointed out by Peter Infante, a large number of the carcinogenic substances identified as threats to the general public have been discovered through investigation of the work environment (Infante, 1995).

If we hear the echoes of lawyers walking in the corridors of power, we should not be surprised. In a system where workers and scientists are wary of one another; where trade unions are not very powerful and, in any event, may have institutional agendas different from workers in a particular setting; where industry places a high priority on profit; and where university administrations are primarily concerned about financial liability, we may begin to understand the preeminence of lawyers and the judicial system. The primary and ultimate rationalization for capitalism is the protection of individual rights: the court system is supposed to give individuals recourse against abuse, negligence and the like. Sadly, in the United States at least, where the labor movement is so weak and the link between public health and labor so tenuous, lawyers end up being the specter that haunts the contemporary work environment.

CHAPTER EIGHT

Work, Health, and Democracy

We began this book with a question: what is the point of production? We started with that question because of its dual meaning. Our focus, like that of Marx and others, remains the point of production: where things are made and lives are lived, structured, and sometimes lost. We began here, also, because we believe that we need to answer such questions are: who is production for? who does it benefit? who controls it? Consider:

- There has been scientific evidence for centuries about the health hazards of lead. So, why do we permit workers to be poisoned by exposure to lead?
- Pesticides are designed to kill pests—but they are toxic to other living things. So, why do we know so little about the human health effects of most pesticides in use today?
- The textile industry has been at the forefront of industrial revolutions throughout the world. So, why did byssinosis, a respiratory disease of cotton mill workers, go unrecognized in the United States until the late 1960s?
- Asphalt fumes have been identified as a carcinogen in Denmark. So, why is it regulated in the United States not as a carcinogen but only as an "air contaminant"?

- A major transnational automobile company has clear internal guidelines for reviewing possible equipment purchases in order to prevent hearing loss among its employees. Why are these guidelines ignored by plant managers?

- Nowadays, there is less full-time work, more temporary work, more work speedup, and more shift work than ever. All of these increase physical and psychological health problems. Although scientific and technical aspects of these developments are well worth studying, are the best solutions likely to be solely scientific and technical?

- When hazardous technologies and hazardous substances find their way into developing economies, after being prohibited in the countries of their origin, what kinds of solutions are available?

- What should workers or poor people or impoverished countries do when confronted with the choice "your job or your health"?

These questions indicate that the structural causes of threats to worker safety are far from solved and are in some cases getting worse. This should not be the case because work is the way in which we shape the world. Work is our creativity, our imagination, our way to demonstrate commitment to community and to the development of our selves. Work feeds our bodies and spirits.

Work should involve risks, but risks to tired minds, challenges to complacency, hazards to inertia. Work must not be the endangerment of some—the workers—for the benefit of the *few*—employers. Indeed, work should also not involve the endangerment of some for the benefit of the *many*—consumers. Democracy, health, and work are inextricably intertwined, not merely in an economic calculus but in one concerning fundamental human value and the rights of individuals and the community.

In the United States, the mass media are filled with news of the fragments of a public health movement. Trade unionists mobilize for health care reform, community groups protest the latest hazardous waste atrocity, the cry for environmental justice is heard in the inner city, people demand the right to know, the right to act, the right to refuse, the right to insist on their rights—even at work. Recently, the public-health struggle against the tobacco companies has risen to new political heights, and consumer organizations are increasingly winning their demands for adequate information on the nutritional and chemical constituents of processed food. All of this seeming street babble is in reality the sound of a bona fide social movement. The noise is from a public health movement that is "beyond fitness," although the broad health concern in the United States certainly nurtures fitness. But what the social movement is truly about is production and work, health and democracy. At present, this set of concerns has no consistent and effective voice.

At the national level nowadays, American politics is socially and intellectually impoverished. In such a world, governmental policies end up being determined more by the financial community's concern over "the deficit" than by any concern over the pain and suffering of the American people. When unemployment goes down, Wall Street frets, over the possibility of inflation. Millions of people—most of them impoverished women and children—are currently being thrown off the welfare rolls without being offered a meaningful way to provide for themselves.

Progressive political forces, including labor, inner-city, and environmental movements, are in relative eclipse in Washington, given the Republican-controlled Congress, and they are fragmented and in disarray in much of the nation as a whole. This dreadful combination is the political context for occupational and environmental health policy right now.

On the other hand, occasionally we see glimpses of the way ahead, of the potential for a newly reconstituted left-oriented public health movement—a time when its fragments might come together in a powerful social and political mosaic. Perhaps we can learn from the occasional triumphs on this front that occur; perhaps we can even learn from our more numerous failures. In any case, we have no choice but to try to learn if we are to avoid what we are seeing around the world: the rise of brown-shirted right-wing "populists"—the Zhirinovskis in Russia, the Le Pens in France, the Buchanans in the United States—all with strong working-class/nativist/nationalist support. The threat of such "populism" may put the brakes on global capital and give some time for the building of a more humane democratic politics. But the right-wing movements that exist certainly erode solidarity and community and bode ill for the future.

The damages already done and the "contradictions" of the present system are revealed by the fact that the world faces continued environmental crisis, a massive problem with hazardous waste, a degraded and dangerous work environment, and greater economic and social inequality in advanced industrial societies such as the United States and Europe. The gap in the standard of living and of health between the developed and the developing world continues to widen.

In Chapter One we defined the work environment very broadly, not only as the place where people work but the point of origin for the environmental problems now confronting the world community. We emphasized that the work environment is and remains about *production*. In Chapter Two, drawing from the essential insights of Marx and his twentieth-century followers, we laid out a "political economy" of the work environment, explicitly linking it to such questions as: who controls the production process and, therefore, the production of disease; who and what controls the definition and recognition of industrial disease and injury; and what the inequalities in power and control have to do with how the problems of the work environment are controlled.

In Chapter Three we discussed the development and characterization of technology. We consciously linked technology and production to globalization of capital and its impact on the work environment. We have asserted that, in practical terms, the political economy of the choice of technology is the most appropriate approach to the analysis of occupational disease and injury. The theme of power, or control, is picked up even more explicitly in Chapter Four. In that chapter we emphasized that ideology and power provide the social and political context for understanding the work environment. It is our experience that this is an area frequently ignored by professionals in the field of workplace health and safety—indeed, we emphasize throughout that the problem of the work environment is a political problem at the deepest level.

The conflict emerging at the point of production is played out between workers and employers in a variety of ways, but it is clear that in advanced liberal democracies much of that conflict is passed on to the state and the legal structures of contemporary capitalism. In Chapter Five we focus on the role played by government regulation in controlling the work environment. While the discussion is centered on the American experience and the sociopolitical impact of OSHA, much of the analysis is appropriate to nearly all advanced industrial societies.

The state's role in mediating conflict between labor and capital in the work environment is nowhere made clearer than in the development of the workers' compensation system. In Chapter Six the discussion of the history and impact of this system emphasizes the way it has been used to remove a major source of conflict and institutionalize it into state regulation. In the United States the workers' compensation system externalizes the costs of worker injuries and provides few benefits to the injured worker.

In the United States the central players in the work environment tend to be the professionals directly engaged in occupational health and safety. These include such practitioners as occupational health nurses, industrial hygienists, and federal and state inspectors. Their training and knowledge about the work environment emerge from a set of social relations structured by classical views of scientific method and the sometimes class-conflicted occupational roles they often inhabit. The science they rely on is often controlled and funded by industry. In Chapter Seven we argue that this underlying bias raises some important questions about the reliability of scientific knowledge and the objectivity of researchers and practitioners.

THE TRIUMPH OF CAPITALISM

We have argued that understanding global economy and technology is central to understanding the work environment. But those changes are also at the root of the destruction of community and of social solidarity. The last vestige of the preindustrial world is the family. And, if we are to judge from statistics on divorce, the feminization of poverty, and the notorious decline of "family values," the family may be on its last legs. This is not to say that the human desire for community and intimacy is gone. Perhaps we can understand the rise of feminism as a response to the dramatic change in gender relations demanded by the expansion of the market into our most intimate social relations. The efforts to reestablish community through sisterhood, coincidental with the mobilization of a female labor force, must be understood as symptomatic of the underlying economic change, as well as good-faith efforts to save a sense of community. Feminism is a struggle to reestablish non-market values in the face of triumphant capital.

The intrusion of the market into family life has had extraordinarily disruptive effects: community values have been smashed into smithereens, transformed into consumerism. The rise of service industries has ripped apart traditional workplaces and benefits accruing to unionized labor. Consumption has become the principal activity—perhaps, the "work"—of much of the population; the provision of services has become its attendant drudgery. The consumer is "creative"; meanwhile the service worker is bored and impoverished.

In response to the decline of quality reproduction, we have the rise of environmentalism, one incarnation of which is a consumer movement embarrassed about drinking champagne from plastic cups. But another more thoughtful and rambunctious environmentalism is based in communities bearing the brunt of toxic wastes, contaminated water supplies, and massive amounts of garbage. These are different movements, the former based in Washington and immersed in the elite politics and culture of conservation, the latter a working-class movement concerned mainly about the fundamentals of survival.

The saddest sight, however, is the trade unions desperate for jobs for their members begging only for the opportunity to do the cleanup work. Workers threatened by layoffs (sometimes actually laid off) from defense plants insist that they be allowed to work at the decontamination of the sites where they have spent their adult lives. Construction workers disagree, insisting that cleanup is the work of their trades (and of their contractor–employer–partners in the business). Minority contractors and civil rights groups contend for the right to clean up heavily polluted urban areas—and indulge in the illusion that "youth" can have a future in asbestos removal, lead paint removal, and hazardous waste work. Only the most sophisticated of the environmental groups and the trade unions are interested in pollution prevention, toxics-use reduction, and clean production. They have penetrated the ideological and material veil of consumerism but are lonely in their heightened consciousness.

Social fragmentation has been the result of triumphant capitalism. In Polanyi's terms, society has become the disconsolate tail of the canine market. But the globalization of capital has also meant the globalization of labor markets. Workers in diverse countries compete against one another for jobs, and workers move from country to country seeking a living. Immigration has become an issue throughout the industrialized world and has spawned nativism on the one hand and identity politics on the other. The notion of class falls by the wayside as nationalism and racism ride on the back of the triumphant market.

The domination of society by the market, however, has had another important effect on community and work. The assault on the public sector—the source of support and security for working people in an industrial society as well as the arena in which working-class organizations can do battle with capital—has been mounted in the name of deficit reduction and financial stability. A society as phenomenally rich as the United States apparently cannot afford to take care of its impoverished citizens. Government's involvement is controlling and regulating insurance, education, and health programs is under attack from the Right. On the other hand, the efforts to rationalize the delivery of medical care in the United States have failed for the time being in a battle against deeply entrenched interests such as insurance companies. Unions in the public sector, the last bastion of trade union strength and an important defender of the public sector, are also under severe attack.

Perhaps a mundane example best illustrates the contradictions that the triumph of capitalism and capitalist ideology presents to the Left. Although trade unions have been in decline, linking labor concerns to environmental issues has at times won some important victories. When the Superfund was reauthorized by Congress in 1986, it included provisions to protect hazardous waste workers, and included funding set aside for a worker training program to be supervised by the National Institute for Environmental Health Sciences (NIEHS). Grants were given to trade unions, universities, and consortia including grassroots health and safety coalitions. These grants provided training, but in the case of the tiny "COSH" groups the funds also enabled them to upgrade their own technical capabilities, to reach out to new groups of workers, and, most important to them, to survive as organizations. Some activists warned that accepting such funding from the government would distort the mission of the COSH groups as advocacy organizations, turning them into staff-driven quasiprofessional groups, as well as making them too greatly dependent on government largesse. The imperatives of financing won the day: the activists were not able to win financial support from the labor movement for the COSHs. OSHA had reduced substantially its New Directions Program. The Superfund program represented some level of survival for the health and safety movement, albeit a strange one: safety training in an industry that did not even possess its own SIC (Standard Industrial Classification) code, while the vast majority of American workers remained in the dark about risks there might be to *their* health. The neo-Maoist ideologues of some COSH groups believed that their politics and cleverness in manipulating funds would triumph over economics, that is, they would figure out some way to continue their battle against sexism, racism, and imperialism while also training cleanup workers and firefighters. Professionals on the Left believed that they could use the funds for training to serve labor and do some good. The COSHs took the money.

Over the succeeding nine years, the right-wing attacks on environmental programs and on government in general has eroded support for the Superfund. Attacks on "regulation" and on trade unions have included the branding of the NIEHS worker training program as a "slush fund" for labor. The manufacturing and construction trade unions continue to do battle with one another, while the service unions have only glimpsed their interest in the problems of health and safety. While the attacks have not been completely successful, funding levels have been cut back, and we are seeing the delicate coalition of professionals and labor severely tested. As some of the COSH groups have come to depend on the NIEHS funds for their very existence, the threat of losing the funding has turned previously successful alliances into studies in bureaucratic back-stabbing.

Thus, the globalization of capital and the triumph of the market have contributed to turning labor into petty bourgeois interest groups and professionals into careerists. Only the hopelessly deluded imagine that there is now a coherent "labor movement," a politically mobilized "community," or, for that matter, serious hope for progressive social change.

BACK TO BASICS:
PRODUCTION AND SUFFERING

Even while postmodernism and the politics of identity create new myths and preoccupations for minorities, workers suffer and the work environment deteriorates. Communities are torn apart by the desperate competition for jobs, housing, and the basic amenities of life. There is nothing "relative" about losing one's hand on the job, nor about the failure of medical services, nor about the material poverty of living on welfare or on the streets. That groups in power have a different "story" from that of the powerless is true, but is only relevant for political analysis, marketing, or, at best, for the building of alliances among the powerless. There is a fundamental truth in suffering that only those completely captured by consumer glitz—or by the craving for money and/or power can miss. Injury and disease in the workplace and in the neighborhood are realities, regardless of who postmodernism–identity politics defines as being in power, the patriarchy or dead white males. Ideology is important but cannot compare in significance to material conditions.

Jobs and work emerge as a common theme in community and labor struggles, although sometimes in a perverse way. In one Massachusetts community, the local environmental organization was unhappy when firefighters complained about a lack of emergency response equipment appropriate for dealing with hazardous waste cleanup activities. They thought the firefighters were holding up the cleanup of a Superfund site. There was no getting them together— both sides pulled in their political allies, to the delight of the responsible parties, who were given one more respite.

The sophistication of community and labor movements leads to a developing concern with the ways and means of production. Though we may not own the kitchen, we have become increasingly interested in the stove and the ventilation unit, the chopping block and the knives, as well as the cuisine. We want sovereignty as citizens and workers as well as choice as consumers. We want at the very least to be present when key decisions about technology are being made. Unfortunately, we are increasingly confronted by "job blackmail." Who makes decisions about technology? The boss makes decisions about technology at the same time he decides about everything else.

Perhaps the most difficult aspect of this situation to understand is why we—workers and citizens—put up with such exploitation and oppression. And here it is that the importance of ideology appears. The rules of the game are stacked in favor of capital, and we, citizens in a democracy, have come to accept the rules. The "manufacturing of consent" has been successful. We have come to believe that we must adjust and adapt to inevitable market forces; the pace of technology is fierce, we must be competitive, we have lost our hold on the political economy, and capital is a wild horse racing through the world while we workers cling to its mane in terror.

IS DEMOCRACY THE ANSWER?

When we began working on this book, we thought of it as part of a larger "democracy project." We thought that the analysis of the political economy of the work environment would help to focus attention on the centrality of production, while avoiding the over-simplifications of contemporary economics. We are asserting the importance of political control over the political economy, and we are old-fashioned in that we hark back to Marx again and again. But we also know the terrible damage to socialism that was rendered by the Stalinism of the Left and the police state capitalism of the Soviet Union. We pursued the notion of "industrial democracy," which we thought to be understandable and sympathetic, although it is also old-fashioned.

Now, five or six years later, we are less ingenuous but more troubled about the democracy project. With the extension of the "free market" throughout eastern Europe and the conversion of China to a sort of military capitalism, and with the failure of socialist and social democratic parties throughout the world, we find that democracy is about property rights. Mystical capital rules, and property has its day. For citizen-consumers, democracy has come to mean choice in the supermarket. We rule through our dollars or rubles or krona, and capital, anxious to please us, ranges throughout the world, finding cheap labor and underpolluted environments to exploit, thereby providing us with cheap goods. College graduates are obliged to clean houses, and the poor in our own country wash the windshields of our cars.

And what as to political democracy? What can political democracy mean in an age of television and mass society? We must distinguish between political parties and candidates committed to submitting to the power of the market. The Left believes there should be a government that helps citizens better adjust to the market; the Right believes that the rich should enjoy themselves and that immigrants should leave. It seems that democracy can exist meaningfully only in a community of shared values. In a fragmented, alienated society, it is merely one more spectator sport.

BACK TO BASICS: FOSTERING A MOVEMENT FOR SOCIAL HEALTH

There is no avoiding the deep contradictions that characterize the politics of global capitalism. The divisions in the working class are profound: they are racial and ethnic, national and hemispheric; they are about gender, colonization, economic development, and underdevelopment. And they are as elemental as who controls which jobs, which communities, which schools; they can even divide people who are of the same race, nationality, and so on. The dreadful competition for livelihood and a livable environment presents formidable problems.

The ambiguous relations among classes add new and even more difficult problems. The creation of a movement for social and occupational health would require tying together the fragments of the working class with middle-strata "professionals," some of whom have conflicted class pressures. But the shifting global economy rearranges the structure of the working class and blurs distinctions between the working and middle class. All of the competitive forces come to bear, and alliances are fragile.

Perhaps the most difficult forces to deal with are the ideological hegemony of the market, consumerism, and individualism. Political discourse is narrow and is dominated by free market myths. The mass media dominate the popular culture and limit the political and social possibilities we can consider. The current low esteem in which "existing socialism" is held forecloses the consideration of humane political alternatives to capitalism. Democracy becomes more and more a thing of the supermarket as politics becomes more and more boring.

Yet, the suffering of the people continues. Production occurs as though workers and communities simply do not matter. The consequences are material: injuries, disease, immiseration, and poverty. These consequences have become a natural result of the spread of global capital, given that its social relations and technologies are so hazardous to human life and environment.

References

Adamic, L. (1963). *Dynamite: The Story of Class Violence in America*. Gloucester, MA: Peter Smith.

Alford, R., & Friedland, R. (1985). *Powers of Theory: Capitalism, the State and Democracy*. New York: Cambridge University Press.

Amott, T. (1993). *Caught in the Crisis: Women and the U.S. Economy Today*. New York: Monthly Review Press.

Appelbaum, E., & Gregory, J. (1988). Union Responses to Contingent Work: Are Win–Win Outcomes Possible? In *Flexible Workstyles: A Look at Contingent Labor*. Washington, DC: Women's Bureau, U.S. Department of Labor.

Ashford, N. (1980). *Crisis in the Workplace: Occupational Disease and Injury*. Cambridge, MA: MIT Press.

Barmann, T. C. (1997, July 14). The Mystery of Occupational Disease. *Providence Journal-Bulletin*.

Barth, P. S., & Hunt, H. A. (1980). *Workers' Compensation and Work-Related Illness and Disease*. Cambridge, MA: MIT Press.

Bayer, R. (ed.). (1988). *The Health and Safety of Workers*. New York: Oxford University Press.

Bell, D. (1973). *The Coming of Postindustrial Society: A Venture in Social Forecasting*. New York: Basic Books.

Berman, D. (1978). *Death on the Job: Occupational Health and Safety Struggles in the United States*. New York: Monthly Review Press.

Berman, D. (1981). Grassroots Coalitions in Health and Safety: The COSH Groups. *Labor Studies Journal*, vol. 6(1).

Boden, L. I. (1995). Workers' Compensation in the United States: High Costs, Low Benefits. *American Review of Public Health*, vol. 16, pp. 180–218.

Boden, L. I. (1979). Cost–Benefit Analysis: Caveat Emptor. *American Journal of Public Health,* vol. 69, p. 1211.

Boyer, R., & Morais, H. (1955). *Labor's Untold Story.* New York: United Electrical Press.

Braverman, H. (1974). *Labor and Monopoly Capital.* New York: Monthly Review Press.

Bright, J. (1985). Does Automation Raise Skill Requirements? *Harvard Business Review,* vol. 36(4).

Brodeur, P. (1974). *Expendable Americans.* New York: Viking.

Bullard, R. (1992). Reviewing the EPA's Draft Environmental Equity Report. *New Solutions,* vol. 3(3), p. 78–86.

Bureau of National Affairs. (1971). *The Job Safety and Health Act of 1970: Text, Analysis and Legislative History.* Washington, DC: Author.

Buroway, M. (1979). *Manufacturing Consent: Changes in the Labor Process under Monopoly Capitalism.* Chicago: University of Chicago Press.

Burton, J. (ed.). (1994). *Workers' Compensation Yearbook.* Horsham, PA: LRP Publications.

Business Wire. (1998, April 8). Worker Injury Biz Booming.

Butler, R. J., Johnson, W. G., & Baldwin, M. L. (1993). Managing Work Disability: Why First Return to Work Is Not a Measure of Success. *Industrial and Labor Relations Review,* vol. 48(3), 452–469.

Byant, W., Zick, D., & Kim, H. (1992). *The Dollar Value of Household Work.* New York: Cornell University Press.

Castleman, B. (1979). The Export of Hazardous Factories to Developing Nations. *International Journal of Health Services,* vol. 9, pp. 569–606.

Chelius, J. R. (1982). The Influence of Workers' Compensation on Safety Incentives. *Industrial and Labor Relations Review,* vol. 35(2), pp. 235–242.

Croyle, J. L. (1978). Industrial Accident Liability Policy in the Early Twentieth Century. *Journal of Legal Studies,* vol. 7(2).

Davis, S., & Schleifer, R. (1997). *Indifference to Safety.* Washington, DC: Farmworker Justice Fund.

Dembe, A. (1996). *Occupation and Disease: How Social Factors Affect the Conception of Work-Related Disorders.* New Haven, CT: Yale University Press.

Donnelly, P. G. (1982). The Origins of the Occupational Safety and Health Act of 1970. *Social Problems,* vol. 30(1).

Donovan, A. (1988). Health and Safety in Underground Coal Mining, 1900–1969: Professional Conduct in a Peripheral Industry. In R. Bayer (Ed.), *The Health and Safety of Workers* (pp. 72–138). New York: Oxford University Press.

Dwyer, T. (1991). *Life and Death at Work: Industrial Accidents as a Case of Socially Produced Error.* New York: Plenum Press.

Eastman, C. (1910). *Work Accidents and the Law.* New York: Chatham.

Economic Indicators. (1999, December 19–January 1). *Economist,* p. 48.

Edwards, P. K. (1981). *Strikes in the United States, 1881–1974.* New York: St. Martin's Press.

References

Elling, R. (1986). *The Struggle for Workers' Health: A Study of Six Industrialized Countries.* New York: Baywood.

Engels, F. (1976). *Anti-Duhring.* New York: International Publishers.

Felt, J. (1965). *Hostages of Fortune: Child Labor Reform in New York State.* Syracuse, NY: Syracuse University Press.

Ferguson, T., & Rogers, J. (1986). *Right Turn: The Decline of the Democrats and the Future of American Politics.* New York: Hill & Wang.

Fox, D., & Stone, J. (1972). Black Lung: Miners' Militancy and Medical Uncertainty, 1968–1972. *Bulletin of the History of Medicine,* vol. 3(2).

Fox, M. J., & Nelson, G. (1972). A Brief History of Safety Legislation and Institutions of the United States and Texas. *Business Studies,* vol. 11(2), pp. 45–59.

Fussell, P. (1992). *Class: A Guide through the American Status System.* Touchstone Books.

Gallie, D. (1978). *In Search of the New Working Class: Automation and Social Integration within the Capitalist Enterprise.* Cambridge, England: Cambridge University Press.

Gersuny, C. (1981). *Work Hazards and Industrial Conflict.* Hanover, NH: University Press of New England.

Ginsburg, R. (1993). Quantitative Risk Assessment and the Illusion of Safety. *New Solutions,* vol. 3(2), pp. 8–15.

Gramsci, A. (1971). *Prison Notebooks.* New York: International Publishers.

Green, M., & Weitzman, N. (eds.). (1981). *Business War on the Law: An Analysis of the Benefits of Federal Health and Safety Enforcement.* Washington, DC: Corporate Accountability Research Group.

Greenbaum, J., Pullman, S., & Szymanski, S. (1985). *Effects of Office Automation on the Public Sector Workforce: A Case Study.* Washington, DC: U.S. Congress, Office of Technology Assessment (contractor report).

Hamilton, A. (1943). *Exploring the Dangerous Trades.* Boston, MA: Little, Brown.

Hartmann, H., Kraut, R., & Tilly, C. (1986). *Computer Chips and Paper Clips: Technology and Women's Employment.* Washington, DC: National Academy Press.

Hirschhorn, L. (1984). *Beyond Mechanization: Work and Technology in a Postindustrial Age.* Cambridge, MA: MIT Press.

Infante, P. (1995). Cancer and Blue-Collar Workers: Who Cares? *New Solutions,* vol. 5(2), pp. 52–57.

Ives, J. (1985). *The Export of Hazard.* London: Routledge & Kegan Paul.

John Gray Institute. (1991). *Managing Workplace Safety and Health.* Beaumont, TX: Lamar University.

Judkins, B. (1986). *We Offer Ourselves as Evidence: Towards Workers' Control of Occupational Health.* New York: Greenwood Press.

Karasek, R., & Theorel, T. (1990). *Healthy Work.* New York: Basic Books.

Kazis, R., & Grossman, R. (1982). *Fear at Work: Job Blackmail, Labor and the Environment.* New York: Pilgrim Press.

References

Kelman, S. (1980). The Occupational Safety and Health Administration. In J. Q. Wilson (Ed.), *The Politics of Regulation*. New York: Basic Books.

Knutson, L. L. (1997, April 9). *U.S. Backs Workers' Restroom Rights*. Associated Press News Report.

Kotin, P., & Gaul, L. A. (1980). Smoking in the Workplace: A Hazard Ignored (Editorial). *American Journal of Public Health*, vol. 70(6), pp. 575–576.

Kuhn, S., & Wooding, J. (1994a). The Changing Structure of Work in the U.S.: Part I— The Impact on Income and Benefits. *New Solutions: A Journal of Environmental and Public Health Policy*, vol. 4(2), pp. 43–56.

Kuhn, S., & Wooding, J. (1994b). The Changing Structure of Work in the U.S.: Part II—The Implications for Health and Welfare. *New Solutions: A Journal of Environmental and Public Health Policy*, vol. 4(4), pp. 21–27.

Lear, W. J. (1992). Health Left. In M. J. Buhle, P. Buhle, & D. Georgakas (Eds.), *Encyclopedia of the American Left* (pp. 301–306). Chicago: University of Illinois Press.

Legge, T. M. (1920). Industrial Diseases under the Mediaeval Trade Guilds. *Journal of Industrial Hygiene*, vol. 1(10), pp. 476–477.

Leigh, J. P., Markowitz, S. B., Fahs, M., Shina, C., & Landrigan, P. J. (1996). *Cost of Occupational Injuries and Illness in 1992*. Final NIOSH Report for Cooperation Agreement with ERC, UGO/CCU902886.

Levenstein, C., & Tuminaro, D. (1992). The Political Economy of Occupational Disease. *New Solutions: A Journal of Environmental and Public Health Policy*, vol. 2(1), pp. 25–34.

Levenstein, C., Mass, W., & Plantemura, D. (1987). Labor and Byssinosis, 1941–1969. In D. Rosner & G. Markowitz (Eds.), *Dying for Work* (pp. 208–223). Bloomington: Indiana University Press.

Levenstein, C., Wooding, J., & Rosenberg, B. (1995). The Social Context of Occupational Health. In B. Levy & D. Wegman (Eds.), *Occupational Health: Recognizing and Preventing Work-Related Disease* (pp. 25–53). Boston: Little, Brown.

Levy, B. S., & Levenstein, C. (eds.). (1990). *Environment and Health in Eastern Europe*. Boston: Management Sciences for Health.

Levy, B. S., & Wegman, D. H. (1988). *Occupational Health* (2nd ed.). Boston: Little, Brown.

Lewis, D. (1992, May 18). Woman Boss Is Harassed, Too. *Boston Globe*, p. 10.

Lubove, R. (1967). Workmen's Compensation and the Prerogatives of Voluntarism. *Labor History*, vol. 8(3).

MacClaury, J. (1981). The Job Safety and Health Law of 1970: Its Passage Was Perilous. *Monthly Labor Review*, vol. 104(3).

Marx, K. (1970). *A Contribution to the Critique of Political Economy*. New York: International Publishers.

McGarity, T. O., & Shapiro, S. A. (1993). *Workers at Risk: The Failed Promise of the Occupational Safety and Health Administration*. New York: Praeger.

References

Mendeloff, J. (1979). *Regulating Safety: An Economic and Political Analysis of Occupational Safety and Health Policy.* Cambridge, MA: MIT Press.

Mintz, B. (1984). *OSHA: History, Law and Politics.* Washington, DC: Bureau of National Affairs.

Murray, T. H. (1988). Regulating Asbestos: Ethics, Politics, and the Values of Science. In R. Bayer (Ed.), *The Health and Safety of Workers* (ch. 6). New York: Oxford University Press.

National Association of Working Women. (1984). *The 9 to 5 National Survey of Women and Stress.* Cleveland, OH: National Association of Working Women.

National Council on Compensation Insurance, Inc. (1994, April). Report of the Fraud Advisory Commission.

Navarro, V. (1982). The Labor Process and Health: An Historical Materialist Interpretation. *International Journal of Health Services,* vol. 12(1), p. 8.

Nelkin, D. S., & Brown, M. (1982). *Workers at Risk: Voices from the Workplace.* Chicago: University of Chicago Press.

Noble, C. (1986). *Liberalism at Work: The Rise and Fall of OSHA.* Philadelphia: Temple University Press.

Noble, C. (1992, February/March). Putting Government to Work on Worker Safety. *Technology Review.*

Noble, D. (1979). *American by Design: Science, Technology, and the Rise of Corporate Capitalism.* New York: Oxford University Press.

Nowotny, H. (1975). Controversies in Science: Remarks on the Different Modes of Production of Knowledge and Their Use. *Zeitschrift für Soziologie,* vol. 4(1), pp. 34–35.

Page, A. J., & O'Brien, M.-W. (1973). *Bitter Wages.* New York: Grossman.

Parker, J., & Solomon, G. (1995). Decades of Deceit: The History of Bay State Smelting. *New Solutions: A Journal of Environmental and Public Health Policy,* vol. 5(3).

Peterson, H. (1989). *Business and Government* (3rd ed.). New York: Harper & Row.

Piven, F. F., & Clowerd, R. A. (1977). *Poor People's Movements: Why They Succeed, How They Fail.* New York: Pantheon.

Polanyi, K. (1977). *The Great Transformation: The Political and Economic Origins of Our Times.* Boston: Beacon Press.

Pozzebon, S. (1993). Medical Cost Containment under Workers' Compensation. *Industrial and Labor Relations Review,* vol. 48(1), pp. 153–167.

Quinn, M., & Buriatti, E. (1991). Women Changing the Times. *New Solutions,* vol. 1(3), pp. 48–56.

Quinn, M., Levenstein, C., & Rest, K. (1996). *Good Practices for Occupational Research in the Private Sector.* Unpublished manuscript, Department of Work Environment, University of Massachusetts, Lowell.

Quinn, R. P., & Shepard, L. (1974). *The 1972 Quality of Employment Survey: Descriptive Statistics with Comparison Data from the 1969–1970 Survey of*

Working Conditions. Ann Arbor: Survey Research Center, Institute for Social Research, University of Michigan.

Reich, R. (1993, May 3). *Speech on 22nd Anniversary of the Signing of the OSHA Act.* Press Associates (Press Release).

Rest, K., Levenstein, C., & Ellenberger, J. (1995). A Call for Worker—Centered Research in Workers' Compensation. *New Solutions,* vol. 5(3), pp. 71–79.

Robinson, J. C. (1991). *Toil and Toxics: Workplace Struggles and Political Strategies for Occupational Health.* Berkeley: University of California Press.

Rosen, G. (1943). *The History of Miners' Diseases.* New York: Schuman's.

Rosenstock, J. (1992). *Education and Empowerment: The Origins and Importance of the COSH Approach to Occupational Safety and Health.* Unpublished manuscript, Brown University School of Medicine, Providence, RI.

Rosner, D., & Markowitz, G. (1984). Safety and Health on the Job as a Class Issue: The Workers' Health Bureau of America in the 1920s. *Science and Society,* vol. 58(4).

Rosner, D., & Markowitz, G. (1987). *Dying for Work: Workers' Safety and Health in Twentieth Century America.* Bloomington: Indiana University Press.

Rosner, D., & Markovitz, G. (1987). Research or Advocacy: Federal Occupational Safety and Health Policy During the New Deal. In D. Rosner & G. Markowitz (Eds.), *Dying for Work* (pp. 83–102). Bloomington: Indiana University Press.

Schor, J. (1991). *The Overworked American.* New York: Basic Books.

Schwartz, R. M. (1993). *The Massachusetts Workers' Compensation Act* (5th ed.). Boston: Feinberg, Charnas, and Schwartz.

Sclove, R. (1995). *Democracy and Technology.* New York: Guilford Press.

Silverstein, M. (1995). Remembering the Past, Acting on the Future. *New Solutions: A Journal of Environmental and Public Health Policy,* vol. 5(4), p. 80.

Smith, B. E. (1981). Black Lung: The Social Production of Disease. *International Journal of Health Services,* vol. 2(3).

Spangler, E. (1992). Sexual Harassment: Labor Relations by Other Means. *New Solutions: A Journal of Occupational and Environmental Health Policy,* vol. 3(1), p. 24.

Street, J. (1992). *Politics and Technology.* New York: Guilford Press.

Szasz, A. (1984). Industrial Resistance to Occupational Health and Safety Legislation: 1971–1981. *Social Problems,* vol. 32(2).

Tait, R. C., Chibnall, J. T., & Richardson, W. D. (1990). Litigation and Employment Status: Effects on Patients with Chronic Pain. *Pain,* vol. 43(1), pp. 37–46.

Tarpinian, G., Tuminaro, D., & Shufro, J. (1997). The Politics of Workers' Compensation in New York State. *New Solutions,* vol. 7(4), pp. 35–45.

Teleky, L. (1948). *A History of Factory and Mine Hygiene.* New York: Columbia University Press.

References

U.S. Congress, Office of Technology Assessment. (1985a). *Automation of America's Offices*. Washington, DC: U.S. Government Printing Office.

U.S. Congress, Office of Technology Assessment. (1985b). *Preventing Illness and Injury in the Workplace* (OTA-H-256). Washington, DC: U.S. Government Printing Office.

U.S. Department of Labor, Occupational Safety and Health Administration. (1996a). *Lost Worktime Injuries and Illness: Characteristics and Resulting Time Away from Work*. Washington, DC: U.S. Government Printing Office.

U.S. Department of Labor, Occupational Safety and Health Administration. (1996b). *National Census of Occupational Injuries*. Washington, DC: U.S. Government Printing Office.

Victor, R. B. (1982). *Workers' Compensation and Workplace Safety: The Nature of Employer Financial Incentives*. Santa Monica, CA: Rand.

Wegman, D., Boden, L. I., & Levenstein, C. (1975). Health Hazard Surveillance by Industrial Workers. *American Journal of Public Health,* vol. 65, pp. 26–30.

Weil, L. (1992). Reforming OSHA: Modest Proposals for Major Change. *New Solutions: A Journal of Occupational and Environmental Health Policy,* vol. 2(4).

Weir, M. A., Orloff, S., & Skocpol, T. (1986). *The Politics of Social Policy in the United States*. Princeton, NJ: Princeton University Press.

Williams, R. (1968). *The Meanings of Work*. In R. Fraser (Ed.), *Work: Twenty Personal Accounts* (pp. 31–46). Harmondsworth, England: Penguin.

Wooding, J. (1990). Dire States: Health and Safety in the Reagan–Thatcher Era. *New Solutions,* vol. 1(2).

Wooding, J., Levenstein, C., & Rosenberg, B. (1997). The Oil, Chemical and Atomic Workers Union: Refining Strategies for Labor. *International Journal of Health Services,* vol. 27(1).

Zuboff, S. (1988). *In the Age of the Smart Machine: The Future of Work and Power*. New York: Basic Books.

Zwerling, C. (1987). Salem Sarcoid: The Origins of Beryllium Disease. In D. Rosner & G. Markowitz (Eds.), *Dying for Work*. Bloomington: Indiana University Press.

Index